THE NERVES AND THEIR ENDINGS

THE NERVES AND THEIR ENDINGS

essays on crisis and response

Jessica Gaitán Johannesson

SCRIBE
Melbourne • London

Scribe Publications
2 John St, Clerkenwell, London, WC1N 2ES, United Kingdom
18–20 Edward St, Brunswick, Victoria 3056, Australia
3754 Pleasant Ave, Suite 100, Minneapolis, Minnesota 55409, USA

Published by Scribe 2022

Typeset in Adobe Caslon Pro by the publishers.

Printed and bound in the UK by CPI Group (UK) Ltd, Croydon CR0 4YY.

Scribe Publications is committed to the sustainable use of natural resources and the use of paper products made responsibly from those resources.

978 1 913348 65 6 (UK edition)
978 1 950354 59 7 (US edition)
978 1 922310 60 6 (Australian edition)
978 1 922586 58 2 (ebook)

Catalogue records for this book are available from the National Library of Australia and the British Library.

scribepublications.co.uk
scribepublications.com
scribepublications.com.au

For Adam,
vars nervios hum alongside mine

CONTENTS

Silba el viento dentro de mí.

Estoy desnudo. Dueño de nada, dueño de nadie, ni siquiera dueño de mis certezas, soy mi cara en el viento, a contraviento, y soy el viento que me golpea la cara.

Eduardo Galeano

When it happens (as if it hasn't already),

we want to make sure we're together. Own enlaced with own: in the right place at this right time. As a time, it has been good to us: to those who count as us.

When this thing hits, we say, we want to make sure we stay *we*, until the end of our tethers —
the borders encapsulating sense. We must hold our nerve,

keep our wits about us. What else is there, but this system of the nerves, and what if it's not about us?

what does it mean to save us, those of us who count

as us every crackling end of us?

'WHAT HAVE I DONE?' AND OTHER ILLUSIONS OF CONTROL

One or two photographs from the winter of 2006 are caged in my external hard drive, carried around between homes, countries, and boxes of life-debris. They are safely stored, always out of the way. One of them is a selfie, from before pictures of oneself were ever referred to as selfies, taken inside a hospital-ward loo. I look stressed in it — I was most likely expecting a carer to knock on the door any minute and ask what I was up to in there. My eyes are large and liquid. I look like an elderly deer (anorexia makes you look both older and younger than you really are, I found — but time is only one of its dislocations). The arm holding the camera is a piece of scaffolding without anything to soften it. At a certain point, I was able to encircle that arm between the tip of my middle finger and my thumb. This was a great achievement — a mastering (not even a squeezing) of the flesh. A few years ago, I showed this picture to Adam, my Person, and realised, once we were both looking at it on the couch of the present, why I wanted him to see it. He's never known me like this, that's why.

Was I? Actually proud. Once, I'd been so very much on top of things.

*

In January that year, aged twenty and thinking birdlike was sexy, but not the least bit interested in sex, I'd been admitted to a closed eating-disorder ward. I spent most of the spring there, re-routed into a strict schedule of meals, and the pooled hours between meals. This was followed by a couple of months as a day patient in the adjacent ward, until I was discharged in the summer of the same year. Throughout this time, I exchanged regular emails with a close friend. Elix was in Malmö, the southern Swedish city I had left in a hurry, no longer able to study, or handle the size of a regular morning. With them in the real world, and me in a place where time was organised by the recurrence of card games and illicit sit-ups, our emails, along with Elix's mix CDs, offered me a line out — a snorkel bringing in sips of air. The messages were a reminder of where I belonged, and an assurance that a space was being held for me there.

My reports to Elix were mostly about the new order of things: the way breakfasts worked (and most often didn't); the staff rotas we used to keep an eagle eye on, anxious to know (for no reason other than simply *knowing*) who'd be working the night shift; the crocodile-shaped key rings I'd learned to make out of tiny beads; how making crocodile-shaped key rings out of tiny beads hadn't been on my list of things to accomplish the year I turned twenty-one. I told Elix about a recent excursion to a farm, where a group of us were taken to visit some heavily pregnant goats, for therapeutic purposes. The patients found it funny — six tiny, anorexic people walking six pregnant goats through at least two feet of snow, with the carers pushing from the back; whose idea was it that this would help with weight gain?

In response to one of those emails, Elix wrote: 'I'm so sorry that you, and everyone in there, have to go through this.' It looked to me as a very odd thing to say.

I admire my friend so very much, especially the wisdom with which they've always, for as long as I've known them, recognised the harm that happens all around us, in spite of us — how difficult it is to inhabit yourself fully and to accept your reach in space. I don't remember addressing their comment then (the emails are long gone now — lost to a defunct email address, something involving a species of aquarium fish); I let it slip and, most likely, continued to regale Elix with tales of the nurse with the torch, with whom I was fighting a nightly battle to keep the door to my hospital room closed. There was something in their message that didn't sit well, though: 'I'm so sorry that you, and everyone in there, *have to go through* this.' The 'have to' didn't seem at all appropriate; it suggested a coercion, pointed a finger to an outside force which, within the gentle, disastrous little universe of the ward, was nowhere to be seen.

Throughout this illness — every small decision dictated by that illness and leading further into it — I would have told anyone that nothing and no one was influencing my not-eating, nor the rabid walking past perplexed acquaintances in Malmö parks (that classmate and his family, saying hi and getting nothing back). I would have told them that I was in charge. Isn't that the most basic definition of self-destructive behaviour? Clearly, it's something you must necessarily do to yourself.

*

Twelve years later, on a Thursday morning, Adam and I begin the day by opening our wee digital windows. What we see leaves us staring out of our actual windows for weeks, eating and communicating badly. The first thing I encounter is a tweet by an artist who says she would 'do anything to save us'. *From which bastard?* I think, lining up world-leading idiots, and then, immediately, there's someone else tweeting: 'this, this is the asteroid'. They mean the one in all the films. That, in turn, leads me to an article in *The New York Times* about the latest IPCC report, which has just been published. It's October 2018. Within fifteen minutes, Adam is watching a lecture about the albedo effect, tipping points, the possibility of a Hothouse Earth scenario (combustible spots on every continent), and the sixth mass extinction. We have moths in our bedroom, carpet moths. When they fly by, Adam shouts 'moth!' as if he were our local town crier.

Within the hour, we no longer inhabit the same world-history as our families or friends. The same week, I snap at a co-worker for using the phrase 'in more mundane matters' when I mention the hell that is materialising, because how could anything be more of the world than the end of it? She is referring to restocking Sellotape. We're in dire need of it in the office.

For those who haven't yet experienced climate collapse in our own bodies, a history not yet written into us, the *feeling* it arrives in the shape of shadows, an atmospheric wrongness, and harrowing predictions; these are stories that change our own. The moment we begin to truly engage with climate science, our narratives of self and future are whirled out of orbit. For me, it was the IPCC report that ultimately tipped knowledge into feeling. The (overly simplistic, it would turn out) headlines declaring that humanity had 'twelve years to limit climate change catastrophe' fulfilled one purpose exceedingly well. They alerted certain groups of people, primarily economically privileged, often white people, mainly in the Global North, of climate collapse as a present tragedy, not a menace a hundred years down the line.

So many of us spot the ghoul of climate change, but only behind half-open doors that we won't walk through, not yet. Not while there's still comfort to be had in 'more mundane matters'. I'd read books, hyperventilated for a while, and then got up to read or watch something else — because what else is there, but the next moment? I spent a weekend handing out flyers back in 2011, when I first moved to Edinburgh, and the next weekend I didn't. I joined Greenpeace, but never went to meetings. Now, environmental collapse forced itself into the same timeline as the next Sellotape order, my next writing project (it wasn't supposed to be this), or my next possible visit to see my grandmother in Colombia, now in her eighties and difficult to understand over the phone since her stroke.

In the weeks after finding out about the IPCC report, different versions of the question 'what have we done' invade my notebook, stuck like dead flies on windowsills. There is 'what have we done?' but also the slightly more inquisitive 'when did we do it?' and lastly 'how come no one is running down the street screaming, if this is what we did?' There is, not least, 'how could we let this happen?'

All of it, really, amounts to howls into cramped spaces. The climate crisis has moved inside, and with it the blame. Out of all directions we could be moving in, this is not the most useful.

*

Considering that blame, it may not be all that surprising that during these first weeks of head-first immersion into climate science, I can't stop thinking about the card game Uno. There was a carer in the hospital, back in 2006, who used to suggest Uno tournaments whenever someone was having a panic attack in the corridor. He was Finnish, very competitive, and had very thick fingers with which he used to hold the cards tightly, a barricade against cheating, coming from all sides. He'd grab attention from a howl outside the communal area by asking you what you'd dreamed that night. In 2018, as Adam and I keep reading (methane beasts, locked-in warming, the Amazon emitting more than it can eat) and start going to meetings, as we train in non-violent direct action, run for the local council, and get arrested for protesting, I remember the people in that ward more clearly than I have in the previous twelve years — the people who, in Elix's words, had to go through it all.

Something in the experience of dissecting climate collapse as an event — something that 'is happening' and you find yourself in, and the attempt to unpick individual agencies in it — feels oddly familiar. It reminds me of arriving in a hospital room on the northern outskirts of Stockholm — on the wall, a watercolour painting of a field with a hat in the middle (no visible head) — and sitting down on the bed to ask my dad 'how did this happen?' ('it's what happens when you don't eat,' he said, checking for plug sockets, which was all anyone could feasibly say). All those moments when I ate less, didn't listen to hunger, went for ill-advised runs piled up over time but under the radar, all that harm being done off-camera and then — in bursts that seemed incredibly rude — coming into view.

Someone tells me that as they've grown increasingly aware of how quickly the climate crisis is escalating, they've also begun to regard their own body as a measuring tool for planetary harm. Vast, intricate mechanisms of destruction enter one nervous system and turn it into the place where global heating seems to begin, where it's perpetuated, and where it worsens. Everything an individual consumes (that banana rather than an innocent turnip), how it gets from one place to another (are you flying home for Christmas?), the space it takes up in the world (are tiny homes and 'micro houses' the answer?) gains significance when you realise — as in, make emotionally real — the connection between your way of life and the risk of societal collapse, that you do not end with your physical boundaries. Your nerves, then, seem to stretch beyond what is visibly yours. Climate scientist Peter Kalmus describes how, since his climate awakening, the thought of flying has given him nightmares: 'It feels like the plane is flying on ground-up babies to me,' he said in an interview.

I want to shout unseemly things at anyone using a plastic bag, or rather, anyone asking me to sell them one (I work in a shop), because they are dragging me into it; they are making me an accomplice in this amorphous crime.

Is it justifiable, that guilt? Perhaps it's necessary. It could be what jolts someone into acknowledging the impact they have on their surroundings. But does it also imply that you have a choice — that you could step outside of all this pain completely, if you only severed yourself from the system, if your choices were perfected, disciplined enough?

*

It's been another couple of years now, countless more re-evaluations. For as long as it doesn't physically crash over you, it always remains easier to look away from the horrors of environmental collapse. No matter the latest hurricane, as long as it's not in your city. Within this state of sleep, I've begun to call those sudden realisations my 'holy shit moments'. Doug, a friend who's a psychotherapist and whom I met through our shared terror of what's coming, has a useful image. He describes the movement back and forth between engaging with the crisis and retreating to ordinary life as a 'misted mirror'. Getting out of the shower, you wipe it and there your face is, warts and all, the lips of someone who doesn't know what 'enough' means. Soon, the mist, which includes your own breath, covers it up again, but every time the mist is cleared, there's the possibility of something new emerging that wasn't there before. Except it isn't new at all; it only happens to be new to you.

'And there are no new pains. We have felt them already,' writes Audre Lorde. My attempts at making sense of the climate crisis have changed and adjusted repeatedly, they continue to shift and to quake. Like so many, I began my response to global existential threat with a focus on raising the alarm, joining efforts to pressure the UK government to immediately reduce carbon emissions (alongside agonising about how we could see my dad and sister if we could no longer fly). The more time I spend with the nervous system of climate collapse, the more restrictive, and counterproductive, this stubborn focus on emissions, and individual emissions at that, appears. 'For 500 years, this has been a place of ruins,' Brazilian journalist Eliane Brum writes about the Amazon. 'The fact that the Amazon is still regarded as something far away, on the periphery of our vision, shows just how stupid white western culture is.' The climate crisis is an illness of severed connections, on a colossal and intimate scale. Those whose connections are thoroughly severed are also the least vulnerable, and those with the most power. Disconnection is both cause and constant generator of mayhem. The Amazon, as Brum points out, is 'the real home of humanity', and it's been the stage of genocide and ecocide for centuries. Colonialism created the climate crisis. White supremacy is its language and logic. In the expansive nervous system of this world, white supremacy, patriarchy, and extractivism are faulty codes, the misfiring signals which disconnect me from what is, in fact, happening to me; they construct peripheries and thereby make ancient pain seem brand-new. As a half-Colombian (the Amazon makes up 40 per cent of Colombia's landmass; the country's history is one written in the blood of European and later US imperialism), as a brown person, I didn't see systemic racism as equally urgent, when I became climate-scared out of my mind. Those weren't the signals fired.

The first time I heard the term 'slow violence' was during a talk delivered by a fellow activist and university lecturer. It was in the upstairs room of a pub in Bath, where we lived then, and the setting could not look safer, or cosier. Bath as a city makes a living out of safe and cosy, as well as quirky and quaint, and as is to be expected, the quaint doesn't fall away even as you're talking about violence. 'Slow?' I said to Paul, the lecturer, because it kept repeating on me; 'slow' next to 'violence' immediately seemed to have a mitigating effect on the 'violence'. It morphed it, turned it into an oxymoron. The next time I saw Paul, he lent me the book he'd been referencing: *Slow Violence and the Environmentalism of the Poor* by Rob Nixon. It's about how the rich wreak havoc on the poor, by the methods through which they got rich. The mainstream media generally tells stories of violence as sudden events that cause immediate pain. There is a cut with the goo and the spillage. If public attention doesn't stick around long enough to witness what comes after the blood has been wiped off the streets, then what about suffering delayed by many decades? Slow violence, I read, 'occurs gradually and out of sight'. This harm 'is typically not viewed as violence at all'. The perpetrator has left the scene. In fact, they may never have, physically, been on the scene at all.

I gawked out a window in that pub in Bath (there's a lot of staring out of windows in this story; it could be that I'm waiting for the windows to cave in), identifying manifestations of normality in everything from a delivery driver to one of the thousands of gulls, whilst sweating into my very regular pint. This was at some point in the summer of 2019. The heat I was feeling was the result of carbon emissions released decades, if not centuries, earlier. That's how long emissions stay in the atmosphere. During one single day that summer, 12.5 billion tons of ice melted from the Greenland ice sheet, into the Atlantic.

Old harm, distant harm. The cut, now the spillage, separated by so many mixed signals.

*

The fashion industry's impact on body image is well documented; it is rarely recognised as violence. The same applies to the effects of diet-related products, or what happens in the wake of certain trends in popular culture. This harm is slow; it occurs across great distances, dissipated by mass communication. Words used to describe this kind of damage tend toward the nebulous (the fashion industry 'contributes' to certain ideals, it has a 'negative influence'), and why not? No one person, or company, sat down one day and decided to conjure up eating disorders. This doesn't mean they've stopped, once they realised the consequences of their work. The violence, by then, is too far diluted, too much at arm's length.

What about a toxic factory being built next to a residential neighbourhood? Who is responsible for the deaths? What about Europe-based companies with investments in the clearing of the Amazon, resulting in the murder and displacement of Indigenous people, diminishing our last chances of survival? What about the ingrained knowledge that I'm worth less than what I produce, especially if I produce less than others, or the pictures of myself I used to prefer — that they were the ones in which my skin looked the lightest? Who is responsible for the hurt that follows, and whose is the profit?

'I'm so sorry that you, and everyone in there, have to go through this,' my friend Elix wrote. Most twenty-year-olds know about the effects of an ideal body image; I knew it too, in the early 2000s. I'm pretty sure, although language mutates and it's easy to apply labels in retrospect, that I thought of myself as a feminist. Two years earlier, I'd been spray-painting on H&M bikini posters at two am in the area around my high school. 'FEJK!!' (fake, but Swedified) my friends and I scrawled across cleavages and Photoshopped waistlines in the subzero night. I was conscious of the effect of nineties midriffs and Weight Watchers. Still, when my friend insinuated that my illness was something done to me, instead of solely done by me, I rejected that version of the story outright. Even when I went to see a trichologist (I'd never heard of a trichologist before, but was referred to one because of thinning hair), even knowing that I still have irregular periods due to those years of being severely underweight, even seeing how that harm is displaced in time, I didn't consider that it could be relegated in other ways. I never allowed for the possibility that my own scalp, looking not unlike an artificially planted forest under the infamous bathroom light bulb, might be a physical result of something beyond myself.

*

If I knew about toxic systems, why did I find it so repulsive to admit that I might have been affected by one? What, if anything, does this say about how people are encouraged to see themselves, as individuals, in a collapsing climate?

The therapist I saw regularly whilst on the ward, a red-haired woman in her thirties of whom my then-partner once said 'of course she wears green', once told me that the entire world is 'eating disordered'. She said it with a patient wave of the hand, as if to say, 'we all know this but it needs repeating.' We were talking about ubiquitous food marketing. In bus stops, in ad breaks, and for over twenty years now, popping up online next to whatever it was you thought you were looking for; food becomes an arena for fantasies of endless choice. These choices define people's lifestyles and, by extension, our sense of self. At some point, something went very wrong, the therapist was saying, because, biologically, we do, obviously, depend on food. Eating is natural: a supposedly good and necessary thing.

What went wrong might have something to do with control, and control seems to lie at the core of both of these stories. In the words of cultural critic Lauren Berlant, food 'is one of the few spaces of controllable, reliable pleasure people have'. It's a jarring word, 'reliable', in the context of liveable habitats going to bits, in the face of societies breaking at the seams. On a global scale, and within an individual body, that order is an illusion. Food is reliable only if you have enough of it. Hunger is controllable only at the expense of your health. Underneath that chimera lies a biological need. To need something — to really require it in a life-or-death way — is to be left utterly defenceless to its loss. To die when it dies.

To some extent, our illusions of control define not only how we eat, manhandle, and love our bodies, but how capitalist society as a whole relates to non-human environments, and those it deems less than human. In one of the first anthologies on climate change and psychology to be published, I read about 'environmental neurosis': the confusion that occurs when people need the rest of existence in order to survive but also can't bear the thought that we're not self-sufficient entities, that the vulnerabilities of ecosystems are our vulnerabilities too. Needing our environment means that we're not in control, after all. This neurosis has many names; another is colonialism. A nervous system, colonised.

Looking at those pictures from 2006, the one in the bathroom and other studies of a peculiar, objectionable, and heightened time, what strikes me most is how I really didn't see my body as part of myself. I was trying to catch it in passing, ring it in, like an unruly flock of sheep. The reason I didn't like what Elix told me was that it robbed me of my fantasy of control, this dogma that 'I chose to become ill', which although absurd was also comforting, because the opposite can look like chaos. Acknowledging any outside influence made me more of a wreck than I already was.

*

'We've emitted as much atmospheric carbon in the past thirty years as we did in the previous two centuries of industrialization,' Jonathan Franzen wrote in *The New Yorker* in September 2019. I have such an issue with this 'we'. Ever since I caught it, most 'we's have, in fact, been ruined for me, because Franzen is by no means alone in his use of it. On the surface, it looks like a shorthand for all of humanity, and without making explicit who you're talking about, it is often understood as such: 'we' are we, the humans, pretty much everyone. The problem being that when it comes to responsibility in destroying the earth, there is no simple, unproblematic 'we'.

As I write this, in media coverage of the Covid-19 pandemic, a similar pattern has emerged, except here, writers are adopting even more overt language of self-harm: 'This is not nature's revenge, we did it to ourselves.' Which 'we'? Who belongs in this 'ourselves' that 'we' did it to? There's plenty of good intentions — a reaching for collective responsibility during a crisis and, as a result, the possibility of collective action. I'm not saying we should never say 'we' anymore; but by remaining unspecific when it comes to agency, 'we' erases the power, the hidden slow violence, that was always at the root of the crisis. It replaces real connections with anonymity. 'We', as it is most often used, are not the oil industry, or the billionaires. When 'we' as a whole are held responsible, patterns of exclusion and oppression continue to strain into the earth, further rooting themselves. The illusion persists.

This 'we' really refers to a small, privileged portion of the earth's population, not its majority. Ultimately, the generalisation is no more helpful than the over-emphasis on individual responsibility: *this is how you lower your carbon footprint*, *ten ways of saving the planet*, Veganuary, and carbon offsetting on flights. Not only do millionaires and powerful polluters have a stake in both simplifications, both extremes, of 'we' (if the solution lies in consumer choices, polluters may still have a role to play, and if a faceless 'we' is the problem, there's no one in particular to hold to account), but they also provide a false autonomy. Real structural violence is complicated, often elusive. It works in layers, from the individual to the companies to the systems above, all ensnared in a hideous dance with each other, all making us very nervous, yet unable to really feel.

Recently, I came across Kevin Anderson's description of his own attitude toward individual versus collective action: 'we have to repeatedly remind ourselves that the separation is nothing but an epistemological construct,' he writes, 'it is not "real".' Maybe what is true about climate action also applies to what caused the crisis. It was never wholly an individual burden, nor the work of an anonymous mob. I do not think I'm writing this because I want to be let off the hook, or because I'm not a part of it, but because this is what it means to be one element in an ecosystem, and to live in a human body — to belong in its demise as well as its flourishing.

*

Call them 'holy shit moments' or a 'misted mirror' — they keep recurring, luckily.

Here's another one: I have never been as cruel to those who loved me as when I was ill, and I have never felt as powerless. I went out for a 'walk-run' (a sneaky run camouflaged as a walk, at a time when runs were strictly prohibited) for hours without my phone, counting ducks and miles round and round a Malmö park and, on my return, found my then-partner dissolved in worry. My dad told me, only recently, that he'd received a call at his office from that partner, who told my dad that he couldn't find me.

I did that to them, and at the same time it happened to us all.

*

When I asked Elix if they thought writing this piece was a bad idea, they said it might feel hudlöst (skinless, a favourite Swedish expression of ours), but that an inability to see the relation between an eating disorder and a collapsing climate is an inability to see the root causes of both. What they really meant in their email that day (I think, I believe, I was unable to see then) when we were so much younger was that perhaps it wasn't all my fault. In spite of the stories we inhale, this is not the same as saying that I wasn't at all responsible.

That illness, a bit like this other, tremendous one, ironed me out between a state of utter culpability and the suspicion that nothing I did mattered — between crushing guilt and total victimhood. It is an isolated, paralysing way to be. It allows terrible things to happen.

*

My more sustained recovery from anorexia — or rather: the beginning of a recovery which happens every time I choose the world, and choose the world again — came not out of mind telling body to obey (*eat! I tell you, eat!*) but with something much more subtle.

After months of eating under supervision, safe structures, and limits, in July 2006 my fellow patients and I were shipped off home for three weeks of summer break, during which the ward would be closed. In my experience, as an eating-disordered person, choosing to look after your own body is the most difficult thing. It means trusting your body and giving it what it needs. A few times, I was tempted to miss meals, to go for longer walks, to say hi to the ducks (different ones, now I'd had to give up my room in Malmö). What kept me resisting was the thought of the others in the ward, and the investment we all had in a vision of getting better, an admission, also, of wanting more, wanting deeper, that wanting, simply, was something we could do. If I began to tighten those slippery reins once more, they would also entrap a girl with parched hands and an even dryer sense of humour with whom I shared a hospital room for fourteen weeks. She also looked older than she was. If I compromised on our new, shared, conception of 'a meal', it would affect the recovery of a nineteen-year-old person with terrible taste in music, and a football fan with a cackling laugh. Some of these people I liked, some I definitely didn't. Knowing them helped.

'To experience solidarity', bell hooks wrote, 'we must have a community of interest, shared beliefs and goals around which to unite'. We had the goal of rejecting an illusion, and tending to life instead. How do we build such solidarity across the immense distances we're looking at — on the scale the present moment demands, is shrieking for, not outside my window but within my body itself — with people whose experiences are vastly different? How do we repair connections, not based on control, but on mutual vulnerability?

Because I do need them to survive, or I won't survive. Because my wellbeing is entangled with theirs, and I need them to get better.

When it happens there

I am training myself in how to feel it here, underneath a left-hand cuticle. Mostly at night I train myself, when what I feel is not what has happened.

signal a butchered response

signal who is holding the axe?

It's the job of nerve endings to pick up harm in the periphery, wherever you come to an end. I get up to piss. My Person feels me leaving, his cochlear nerve tuned past our first languages. There's a wave of wildfires moving through Siberia.

In such a knotted system

what counts as a periphery?

(Has anyone ever hit a nerve — punched it, and watched as it fell?)

When is on touching terms with *where* only if you shut your eyes or open your nerves wider, crack those bastards ajar at the rim. Could you burn the living ends through their cotton wool coatings.

I don't mean about being raw,

but about *rawing*.

THE WAYS WE USED TO TRAVEL

Min far sa: Den som reser är överflödig för platsen hon kom ifrån
Min mor sa: Den som reser tror att hon är oumbärlig för platsen hon kommer till
Min far sa: Den som reser är överflödig för platsen hon kommer till
Min mor sa: Den som reser tror att hon är oumbärlig för platsen hon kom ifrån
Min morbror sa: Den som reser vet ingenting om plats

My father said: The one who travels is redundant to the place they came from
My mother said: The one who travels thinks they are essential to the place they come to
My father said: The one who travels is redundant to the place they come to
My mother said: The one who travels thinks they were essential to the place they came from
My uncle said: The one who travels knows nothing about place.

Athena Farrokhzad, translation by Jennifer Hayashida

Hyphens

Somewhere along the way, I became neglectful of distances. Because it was never my own knees giving in, my own back paying for passage; the weight of miles became an afterthought. You stumble upon strawberries in your local supermarket in January and you say, 'maybe I fancy some.' It shouldn't be this easy, but you plonk down a punnet and they look *like* strawberries. Strawberries in January. Artificial speed, compressed distance, is also what it takes to get home, once you've decided to leave. This strawberry got here before going mouldy. I'm like a yoyo and, until recently, I didn't ever get very dizzy.

On the map, a straight line between the south of England and the south of Sweden looks like nothing more than a hyphen between longer journeys, as in more noteworthy routes, in the grand scale of things. Both countries are in the northern hemisphere, with a relatively narrow sea between them (one which the Norse crossed over a thousand years ago, without combustion engines). Denmark is in between too, but by the time you get to Denmark, you're practically there. I used to fly back to Sweden from the UK two or three times a year. The speed of commercial flight in the early twenty-first century makes it possible to cross that distance in roughly two and a half hours, about the length of many feature films. As the crow flies, with the speed available to people like me, you hardly have time to even notice Denmark.

In the summer of 2019, Adam and I decide to go overland to visit my family in Sweden. We've decided to 'stay grounded', as the hashtag has it, but we won't call it that, because it is a hashtag. Instead of two and a half hours of flight, it will take fifty-six hours from Bath to our final destination, a small red house with blue window frames by lake

Verveln, 174 miles south of Stockholm, where my dad drowns mice in the basement and builds birdhouses above ground, where my mother and I tried to grow sugar snaps, and she taught Adam how to make cinnamon buns. This includes a twenty-four-hour stop in Amsterdam. We worry that if we don't stop along the way, if we don't notice if a Dutch field is different from a Belgian one, if we don't annotate the weather, we might still miss something essential about the distance — the whole thing might not be grounded enough. It might stay a hyphen.

I do notice the weather. It's like someone else's baby crying; we all hear it, but it's no one's business to bring it up, to see to it, to fix what is obviously in trouble.

Exceptions

Starting with a coach ride from Bath to London, at three am on a Monday morning, the weather is still asleep. Adam was supposed to have time to shave. On the tube between Paddington and St Pancras, I spot him looking for something between all the heads, the bags, and the pre-emptively bare shoulders (it will be hot, it will get hotter), but it's not a space to sit. He's after the one person who's not on their phone. At home, sometimes I'll be swiping at my screen for too long and when I look up, he's just there, staring at me with his mouth in a petrified 'o', like he's been stopped in the middle of a sentence and left there, a line gone mute. It's not a particularly nice thing to have done to someone. We call this look his 'phone face', and I really am trying to do better. We're too old for our backpacks, I think. Then, just as I'm about to get my phone out, he looks up, triumphantly. Someone, squeezed in on the right side of the aisle, is reading a book.

By Wednesday afternoon, we'll find ourselves in the house by the lake. Perhaps because we're following railway-tendrils, instead of hopping into that compressor of time and space, the commercial aeroplane, we might not even need to lose ourselves at all. There will be no abrupt changes, only gradual adaptation.

The cottage itself is called Kruthult, 'Gunpowder Cottage'. We don't know why, but it was built by the father of someone who now spends their summers nearby. There's no history of an explosion, only railway tracks, toe-biting crayfish, and a misanthrope farmer who chooses to forget about the Swedish right to roam in order to guard *his* mushroom patches, the same patches my dad has mentally mapped with military precision. In the house up on the cliff, around the bend from ours and looking over the lake, someone took their own life. My mother tried to establish the nickname La Ponderosa for the cottage, from the TV series *Bonanza*, when my parents bought it in 2002, I think because she felt that Spanish needed help to hold its own in our family. It didn't stick. We did speak some Spanish there, as we did everywhere else. It's been two years since I was last at Kruthult. It's never taken me this long to get there, and to get all the way there.

We exit London just in time before the heat reaches, like a leak from neighbour to neighbour, all the way underground. It's, as they call it, exceptionally hot.

Hyphenated

I learned how to travel from my parents. When my sister and I were kids, my mum showed us how to roll socks into the gaping mouths of shoes, a space which was also good for storing shampoo bottles, with any excess air squeezed out of them, so as to avoid miniature explosions in the hold of the plane. A few days before departure, we were instructed to put aside the clothes we planned to take with us. We learned that you shouldn't cover the bottom of a suitcase in books, as this gives the impression of a secret compartment when scanned through customs, in particular when leaving Colombia during the mid to late 1990s. Avocados were fine to smuggle as long as they'd been checked in. I once took one home to an ex-partner, then left it with my parents by mistake. My mother posted it from Stockholm to the village in the south of Sweden where I was studying. It was perfectly ripe by the time it arrived, handed over by a bemused receptionist at my college. The boyfriend attempted to grow an avocado plant with it, but it never took root. He'd never seen one that huge, so unlike any avocado he'd ever known.

So much of who my sister and I are, and with whom, depends on the ability to move easily between countries, between identities. My Swedish dad went to work in Bogotá in 1980, where he met my mother, who was then at university, studying economics. They moved to Stockholm in 1983. We were all living in Ecuador when my sister was born, but flew to Bogotá for the birth. The hospitals were better there, I'm told, and my grandfather was a surgeon in one of them. Some places are safe enough to live in, for some people, but not to give birth in, if you have the privilege to make that distinction. After we moved back to Sweden, we used to visit my grandmother, uncles, and cousin in Bogotá once a year, spending about a month there each time. In the mid-nineties, we lived there for three years, and my sister and I went to a bilingual — Spanish and English — school. I lined my throat with a Colombian accent which hasn't been chafed off yet, even though

I haven't lived there for twenty-five years. Sometimes, tourists from Spain are funny about it. It gives me a taste of new racisms. We moved back to Sweden on New Year's Eve 1996. My sister eventually left to study at a London university and remained in the UK for nine years. I moved to Edinburgh almost a decade ago, and met Adam, who's from Yorkshire. About five years ago, my sister moved back to Stockholm, carrying a small Welsh dog, drugged on tranquilisers, as hand luggage. This was before Brexit, when such a move was still possible.

That dog is Swedish-Welsh now, as we are Swedish-Colombians. Swedish-Colombian-Scottish, in my case. The hyphens, those identity bridges, expand to so much: nationalities, ethnicities, as well as the queerness it took me so long to even begin to hyphenate. In the UK, a double-barrelled name is often equated with poshness, whereas in Colombia it is customary to use the name of your father followed by your mother's name, both constituting your surname, no hyphen needed. I got used to this when living in Colombia as a child and called myself Jessica Johannesson Gaitán for over twenty years, but my passport didn't agree, as Sweden didn't use double surnames, and they'd included Gaitán as a middle name. Eventually, I relented and changed. How many hyphens are enough to feel whole? I'm not posh, I want to say, just half-Colombian, but it would be unwise; I'd be making excuses for much more than double barrels. The term, incidentally, originally referred to the barrels of guns.

Non-Places

The check-in for the Eurostar looks like a thousand of its airport cousins. Although this time we're diving into the earth, instead of leaving its surface, even the chairs you wait in are familiar: the tired slipperiness of them, and the way I expect them to smell like smoke, even though no one has smoked at an airport gate for a long time. This is familiar, as in: it is to do with family. My cousin and I used to run toward each other, a pair of small, unhinged wrestlers, across the arrival terminal at El Dorado airport, year after year until we were teenagers and too far gone for that kind of unrepentant tenderness. Luggage-reclaim halls still make me giddy, even when there is no one waiting on the other side of the doors.

At one point during university, I came across Marc Augé's writing on the 'non-place' and immediately started salivating. I was trying to hone down a dissertation topic, and finding it impossible to choose between places, or even national literatures. This, like the whole scope of my MSc programme (Literature and Transatlanticism — a programme at Edinburgh which no longer exists), seemed like a way to explore belonging without choosing a side of the Atlantic. A whole range of non-place-related stories followed Augé's book, some lifelong buddies, such as Jamaica Kincaid's *A Small Place*, and some that didn't stick around for long, including Alain de Botton's reflections on his time as a writer-in-residence at Heathrow airport. I watched the videos of him sitting at a desk at terminal five, and started collecting scenes set in airports in different novels, thinking I could stay where I was, in the middle.

'What's this nonsense about airports?' my tutor said during our next meeting. It was a bit like when I began to burn incense every day, during my teens, only to find out it was giving my mother asthma attacks. So much makes perfect sense, seems thrilling even, before you find out that it is taking away someone's ability to take a proper breath.

April 2005 was the last time I went to Colombia with my mother. It was the only time we went just the two of us, and I had a fever on the way back. As I lay stretched out on a bench during the lay-over at Charles de Gaulle, counting passing varicose veins, we spoke about sex and exercise, although some time must have passed between the two topics. I find swollen feet, the kind that transatlantic flights give you, almost nostalgic, which, in turn, really bothers me, because by the time my mother died, she also had swollen feet.

As for the sex, I didn't tell her that at the time I was in love with two people, a boy and a girl, although I think she suspected. I never came out to her, in so many words. My internal maps didn't firm up enough, in the years we had left, to be able to tell her that my wanting for others, between sexual and platonic, regardless of gender, is another way in which I have always been in-between.

I recently re-read sections of Augé's book on the 'non-place'. Even though it's short, I had real trouble finding whatever bits seemed so hot back then. In fact, it turns out that I'd misunderstood most of the book's central premise. To Augé, the 'non-place' is the opposite of an 'anthropological place', by which he means 'relational, historical and concerned with identity'. Lacking all those things, he argues, the traveller's space (including trains, aeroplanes, and all kinds of stations) is the archetype of the non-place.

So, what about the tumbling cousins, what about the sex-talk with my mother? Relational, concerned with identity.

When I was five, my dad leaned closer to the window on a flight between Quito and Bogotá, to show me the Andes during an unusually clear night. The snow looked like mould growing in stains on the dark rock. We were the ones leaving, then, not the snow.

Stranded

In Amsterdam everyone's faces seem accelerated by the heat, every single person five minutes late, from wherever, to something else. The heat seems always to be treated as accidental, an awkward scene taking place next door which has nothing to do with the task at hand. Adam and I watch Dutch ducks circling each other, one trying to catch a glimpse, or find shelter, under the other's skirt. I make him stand next to a building saying 'I am A'DAM' so that I can take a picture, because there are things we do know.

Surveys have showed the people of Osnabrück to be among the most satisfied in Germany, according to Wikitravel. Åsna is donkey in Swedish and brücke is bridge in German; Osnabrück sounds like donkeys crossing a bridge, for ever and ever. The wait in this German town is supposed to be an hour, but the next train is delayed by seventy minutes. This means that we'll miss our night train from Hamburg to Copenhagen. The next one isn't till late in the morning the following day.

Even though the places we call home have stayed still on the map, the journey between them no longer feels like a straight line. The crow that flies here and there hasn't been a bird for as long as I've been alive, but a fleet of machines, spewing thousands of tons of CO_2 into the atmosphere every year — creating this heat that keeps rising, after sundown — available only to a fraction of the earth's population, 11 per cent in 2018. The privilege of fast travel hasn't only helped create the climate crisis; it has changed, distorted, the very reality of distances, by widening the chasm between the one who has access across them and the one who does not.

The house by the lake was chocolate brown when my parents bought it, but the first summer we painted it red. My mother never liked the atmosphere of old houses, but old houses in Colombia were far worse than old houses in Sweden. In a former life, she said, she was either a slave in the house of some Spanish aristocrat, or a pig sent to slaughter, because she also never liked lechona, a Colombian stuffed-pig dish. The house rests on a grassy elevation, which gives it a view over a small pond, and the lake beyond it, where everyone swims, except for this one neighbour, who died a few years back. He always chose the pond. We used to see his pale backside between the leaves, taking careful, perfected, steps.

The guy at the information desk tells us to head for Hamburg anyway when we can, and to ask for a hotel voucher, but he doesn't say in what order these things should happen. In the half-deserted entrance lobby, we cool down with ice lollies and regret engaging in conversation with a man who tells us, trying to kiss my hand, that 'we're all Germans'. Back on the platform, it's just gone eleven; pockets of air are released between our movements. The tarmac gives in to the strain of the day, finally allowed to exhale.

How is this actually *felt*? What is getting in the way of us feeling it?

'What do you think you can do,' I remember my dad saying a few months ago, 'just you, with Trump in the White House?' Not much, is the answer, but everyone, everything, responds, and I have to live with my responses. There's the distance between Adam's feet and mine. There's no reason for me to cross it just now, but I do it because I can, to close the smallest gap when confronted with enormous ones. What distance is reasonable, responsible, and what speed?

Synapses

On another German train, the one between Osnabrück and Hamburg, Adam reads and reads, whilst I try to but, instead, think about how I don't know what kind of metal this thing is made of. There are two Swedish boys in their early twenties in the seats behind us. It used to be the same when I was young and got on planes to Stockholm: the first conversation heard, in the language I was heading toward, was an announcement of what was to come, the point where one part of you became another, a translation took place.

At Hamburg Hauptbahnhof, about a dozen passengers gather at Deutsche Bahn's information desk. It's one o'clock in the morning and half of the queue consists of Swedish women in their fifties and sixties. Swedish is quickly taking over; one step through a ticketed door and then it's the norm. The seventies was the golden age of interrailing. 'Maybe they never stopped,' I whisper to Adam. Unexpectedly, we are shepherded into groups, led out to the car park outside the station, and tucked into seven-seat taxis. We end up sharing one with three Dutch teenagers, all of them stone quiet as soon as we get into the car. That carbon footprint, I think, it's fucked, and then I fall asleep. A while later, a vicious sideways jerk wakes me up. 'Is he dozing?' I hiss. I can only see the back of the driver's head, stern and upright. 'It was a deer,' Adam whispers.

This kind of thing is happening now, apparently. I was never afraid of flying either, but when something is so fundamentally wrong, very little can be considered safe, balancing on top of it.

Something else I hadn't paid attention to, back when I was reading about the 'non-place', was how little Augé mentions its impact on the patch of earth, the ecosystems, the real oozing or desiccated place on which non-places were erected, and from which they suck their profit. The book doesn't give space to the people who, day in and day out, clean the loos, guard the doors, serve the coffee, keep the non-place going — those whose everyday lives and toil facilitate what others call breaks. Is calling something a 'non-place' saying there's no one there, or that it's not, really, on earth?

This, a connecting water between shores, is a place as well as any other: two or three o'clock in the morning on the upper deck of a ferry, crossing the sea between Germany and Denmark, with the wind prying clothes open, sneaking into the envelope between lid and eye. There are fish who live, procreate, and are familiar with each other, down there. Proper night also brings solace from the heat. Once in every twenty-four hours, it takes a deep breath and you can feel your periphery once more. The temperature of my blood is remarkably warmer than the air outside, and the weather has my back again, won't give in as easily to the expansion of my blood, until morning. In Chennai, India, temperatures have reached 50 °C this week. They don't have nearly enough water.

Since I was twenty years old, I've been taking medicine for anxiety on and off. It's a 'selective serotonin reuptake inhibitor', a very common antidepressant, which makes it so that serotonin — the neurotransmitter that regulates things as varied as mood, cognition, and vomiting — stays between nerve cells for longer, doing its thing to make me fit for purpose. If chemical synapses, the space between nerves, weren't that significant, if the connections weren't as much a part of the body as the elements on either side, then how come they have such power over me?

Carbon emissions through transport don't occur inside national borders. Governments wash their hands of it by labelling the places between countries as no places at all, yet through those interludes a capitalist system emerged; by way of those synapses, the same governments live, and make their living.

Severed

My dad has cut down a tree, and my sister has taught the dog not to run for the neighbour's front door as soon as he's let off his lead. He likes to sit there, on their porch, cocking his head and expecting a splendid welcome.

A few times a day during the summer months, a train called Kustpilen, 'The Coastal Arrow', sweeps past on a raised track which divides the pond from the lake. There's an old train station which was closed down at some point in the nineties. The nearest working train station is Kisa, about half an hour's drive away. When my mother was trying to teach me how to drive, we used the first stretch of that journey, through the woods, for practice. It went remarkably well dozens of times, and then, during one short trip, we met four tractors, one after the other. The local train station is now used, along with so many other houses around here, as a summer residence. A cardiologist stays there, who sometimes invites my dad over for dinner. Today, my dad picks us up in Linköping, the nearest city, as I never did learn how to drive.

'The water is cold,' I note. 'No,' my dad says, which might be objectively true, but he's always slept with the window open, all through winter, forcing my mother to wear layers. On the first night, we go down to the lake after dinner. I walk out on the pier and land face down at the very edge, closing one eye to the tufty hairdos of the pines on the opposite shore, and listen for loons. Loons are probably the only bird whose call I'd be able to identify, without the trace of a doubt. To me, the water, the cliffs, and wild strawberries here all sound like loons, because there's no splitting one presence from the other. The area close to the pier, now metallic and inscrutable as the sun recedes, is where my sister and I used to wave at the train passengers rushing past. We always hoped we looked older than we were, jumping up and down in our bikinis; then we'd plunge underwater and shout 'banana' at each other through the verdant dark.

Last time I was here, this place still held my mother in it. The water remained a substance in touch with her body, a body itself that she breathed over, belly-laughed in, and was upset when she couldn't get into, due to her disability. It's a lake around which she was considered a foreigner and — for much of the last thirteen years — felt a foreigner in her own body, but which she left particles, language, and choices all around. It still is that place. Death doesn't remove a person from a place; if anything, it blurs the boundaries between place and person. The problem, perhaps, is the rest of us.

About a year after this trip, I will read Natalie Diaz's poem 'The First Water Is the Body'. Diaz, a Mojave poet and enrolled member of the Gila River Indian Tribe of Arizona, writes: 'If I was created to hold the Colorado River, to carry its rushing inside me, if the very shape of my throat, of my thighs is for wetness, how can I say who I am if the river is gone?' How is my mother in this water, and who are we, who have moved so much between waters? Elsewhere in the same poem, Diaz writes about how this image of body and water being the same lends itself in 'American imaginations' to 'surrealism or magical realism'. This makes me laugh, because when thinking of Colombia, my mother's land — my motherland — many, not just Americans, also think about magical realism. It may be easier that way.

I know, of course, that a thousand worldly fears are exacerbated by personal grief. This is not, by any means, an explanation of what I'm terrified of. It is as simple and as entangled as this: my mother taught me how to live between places. The year she died, I also happened to realise that I knew so very little about them.

Plunge

There's an abundance of snakes this year. They're small but still news. After crossing the railway tracks, we follow a short footpath between bushes to get to the lake itself, stomping through with a menace for the sake of the dog's small nose, and what might bite it to shreds. 'Have you heard of the parakeets in London?' I say to Adam, and then I can't remember the rest of that story, only that it's about another adaptation. I speak to my uncle in Bogotá, who tells me that he hasn't seen a butterfly in months.

The term 'solastalgia' was coined by philosopher Glenn Albrecht in 2005. It's a hand-holding between the Latin 'solacium', comfort, and the Greek 'algia', pain. A pain brought on by comfort, or a pain where comfort should have been? Albrecht himself defined it as 'when your endemic sense of place is being violated'. He often came into contact with people in distress because of the effects of mining upon their communities in Australia. It could be a condition of more than one, mutually enforcing, strand: the discovery that someone's home has changed forever, and the pain of the diagnosis before the full-scale blow.

We each have individual ways of getting into the lake. My dad and my sister never divert from their tried and tested methods. She's a dancer and has gone back to university to become a physiotherapist, which gives her three years of summers as a student. She's spending them mostly here, with my dad, where they've developed routines of their own — half a family of four. When it's time to swim, my dad steps down from the smaller pier and walks steadily out, gauging temperature, checking clouds, saying 'umpf' a few times as his shape cuts through the surface. Then, he turns 180 degrees to face the shore and, expanding lungs and midriff with sealed jaws, collapses backward into the water with something life an 'ujujjuuuj' trailing at his sides, along with shoals of minute fish. The loons are used to it and don't fret. My sister has a different approach. She walks, with purpose, from the grass-lined bank, slowly into the water. It's a tactic for true masochists. Once the water reaches her stomach, she launches nose down without a moment's hesitation and swims a few strokes, before resurfacing somewhere else. My own routine used to involve a jump or a dive from the pier. It would be a good day, when it started head first. Adam is developing his own strategy. 'I'm going to Norway!' he says, side-stroking. 'You're going in the wrong direction,' I say.

During this five-day visit, I don't jump in a single time. The water is warmer than it's ever been, as far as I can remember and according to my dad's observations, but I linger on the steps, stalling and staring at the way light clusters, imagining what the liquid cut will feel like, the immediacy of the surface when it's broken and no part of my skin is safe. It will be expecting the cold, the nerves reaching out to attention, but they will never be ready. I never know how they'll respond, only that they will. It will never be the same. At least one person, perhaps two, around this lake must be frying eggs, I think, and step back and forth on the pier, like a cold-footed mallard, stuck between now and after. I have become suspicious of every sudden change, a state overturned by another before you know it, which is something I should be used to. It was so much like travelling.

In Touch

'If we lived here,' I say to Adam, sitting on the porch with my sister's dog (his name is Arnold but we refer to him as Grisen, the Pig, or Lilleman, Little Man, because there are no grandchildren) belly-up on my lap, 'we could come here all the time.' 'We don't drive,' Adam says. 'If we lived here,' I say to Arnold, 'these naps together would be on the cards much more regularly. I could look after you,' referring to when my sister and her boyfriend need dog-sitting. What I really mean is that I could look after all of them better. I mean that moving back permanently should be my response to all of this — that staying in one place is the only way I can, really, take responsibility.

It's one of the wicked aspects of the climate crisis: addressing it demands thinking globally and living locally at the same time, connecting realities across vast distances whilst, crucially, becoming more of what we always were: part of a physical place. Within this network into which I was born, how do I best look after them: by being close to them, or fighting the arseholes who think they can get away with repeated genocides? Tying myself to a Scottish oil rig? With a crisis this big, where am I most of use, and not just to these specific people, whom I happen to love?

It's an odd thing to feel homesick for a place where you've never actually lived, only lived in the breaks between lives. So few people really live here anymore; they've all moved to the cities. The train station and the post office are gone, and big chunks of the woods have been sold off by the aforementioned farmer, who lives in a house we call Sockerbiten, the sugar cube. It looks like it could harbour a meth lab. 'It's hard,' George Monbiot wrote back in 2006, 'to contemplate a world in which our own freedoms are curtailed, especially the freedoms which helped shape us.' He was talking about the very essence of privilege — the blindness it makes possible. How hard is it, really, to contemplate a world in which billions aren't murdered for profit? The reason that it feels hard is because identity is where we operate from, how we exist, the 'us' that was shaped over time and through love. When my sister and I left to study in other countries, my mother said that they — by which she meant her and my dad — only had themselves to blame. They'd raised us to think that was a thing people did, if they could. She didn't say it like that; she usually switched languages in the middle of a sentence, and she certainly wasn't speaking English when she said it. La culpa es nuestra, she said. Our hybridity never stopped with place: it has come to inform so much, every kind of being and wanting I do. Including my fear at this moment, my responses to a global crisis.

Every day for many years, my mother called her mother in Colombia to play some music and read some prayers. My mother wasn't religious, but my grandmother is, and she'd had a stroke, so God was of some help as an interpreter. When my mother left Colombia in the 1980s, they were only able to speak on the phone once a month; the rest was handwritten letters, which took about a month to arrive. Hers was a very different migration compared to mine — ours is light years away from that of most people who are forced to leave, and who are kept from arriving. In the 1990s, environmental philosopher Henry Shue famously made the distinction between subsistence emissions and luxury emissions. The ability to visit loved ones, with relative ease across long distances, may be necessary for *you* to stay *you*, but this is different from staying alive. If we're able to stay in touch, no matter what, this makes us forget that it all depends on what we can actually touch — the dirt, the water, and the bodies, her sugar snaps and her finger nails. In this crisis of disconnection, how do we repair those most essential links that nourish empathy and multiple perspectives, that make us hybrids, while understanding that many of them were always a luxury, always an exception?

Contrails

We're on our way back, to work, to meetings and a big protest in October, to looking after the neighbour's giant cat, who climbs our bookshelves and the back of my dressing gown. Taking the train both ways is more than we can afford, in annual leave as well as wages. Flying one way rather than both feels better, but like slapping someone's face instead of kneeing them in the crotch might feel better. On the plane back to Bristol, a young person is humming to the music in their earphones. They're soon told off by a gentle but strong-jawed air steward, and this is followed by the signal for strapping on seat belts. The sound is known, no longer familiar.

We were only away for five days. Next time, we say, we'll plan it better, we'll book way in advance and make sure we stay for longer, to make the long journey there and back worth it. We'll adapt to true distance. We're amazed at the quiet toddler in the seat in front. Children so far above the ground are somehow an even stranger thing compared to adults. Other aircraft sew zips of vapour across the sky. They leave temporary zigzag scars across our aeroplane window, one dissolving into the other. Contrails, they're called, trails of clouds. Very soon, we won't see them and there they will be, reflecting more heat back to earth than the rest of the sky.

Postscript

That was the summer of 2019. Finishing this piece, in the summer of 2021, I still haven't been back to Sweden. I speak to them every day.

My mother said that when her favourite aunt died, when the youngest, coolest of all her aunts died, she dreamed the dying. She was in her teenage bed, in the old house in Bogotá, and she witnessed her aunt rising.

Past the mourners, past the doctors and the tubes around a sterile bed, a clinical non-place, where no one would ever make history.

She wasn't waving as she rose but there was a spring — not in her step, but in the cloudless rising. And my mother's sleep replied. She wasn't surprised when they woke her, to interrupt this neural dialogue, to jab her first-hand knowledge with the news.

What kind of signal is this, not seen but often seen as obscene? They'd call it magical realism, especially if it happened in Colombia. They'd file it as a cultural thing.

It was just a synapse exploding at the close.

THE GREAT MOOSE MIGRATION

Few animals are more Swedish than the moose. We grow up with their antlers poking down into our cradles, passing unbothered in the background of a back-seat car window, and then gone. But here's something I find confusing: an elk and a moose are both different and the same, depending on where in the English-speaking universe you find yourself when you see one. What British people call an elk is an American moose, the one with the droopy nose and smooth, plump antlers (as if they've been inflated from the cranium, straight into a souvenir shop). But there's also an American elk: a stag but with more muscle, pointier head, and antlers. The American elk, from what I gather, is closer to a Swedish hjort. The Swedish älg, what my mother called un alce, is therefore both a moose and an elk. It's such a mythical creature, not as in imagined, but as in shedding myths as it goes, with every passing season.

*

For present purposes, I will stick to 'moose'. There's no point in pretending that most Swedish people don't get the bulk of their English, its rhyme and its reason, from American TV shows, even if I've spent years trying to lose the unwarranted accent.

Slow TV Day 6: As of yet, no moose are there to be seen. For the first ten days of broadcasting, the reindeer are the undisputed stars of *Den stora älgvandringen*, The Great Moose Migration, on Swedish web-TV. At half ten in the morning, eight reindeer are resting in the snow between water and wood. Seven of them lie in a half-moon, with their white behinds to the camera, whilst an eighth is holding court in front of the group. Sometimes, there's a shiver, as if they were one body, given a little electric shock. Someone told me, when we moved to Bath, that Nicholas Cage owned a house somewhere in the area. There was a rumour that he showed up in a pub on New Year's Eve and bought everyone a round. At this early stage, the reindeer are the Nicholas Cages of The Great Moose Migration: the kind of celebrity you might, realistically, one day see.

My dad mentions the show a few times before I decide to tune in. By then, the network is almost a week into the live streaming, we are two weeks into the first UK lockdown, and it's been three weeks since my dad last touched another person. At this point, this feels like a long time. My sister is doing his shopping, which he's not happy with, as she replaces regular crisps with vegetable ones. Really, he wants to write the list, pick up the things, bring them home, and thereby plan his week. He's staying indoors in the south of Stockholm, and Adam and I — on furlough from our retail jobs — are staying indoors in Bath, but in the cities themselves conditions are different. The Swedish response to the pandemic is markedly less robust, compared to the rest of Europe. 'People in Sweden have a high level of trust in government agencies,' the Swedish prime minister says in an announcement, reminding us of who we are. I'm not sure I count, having left Sweden almost a decade ago, having always been something else too. I feel like he's speaking to certain trace substances. As a result, he adds, 'people in Sweden are on the whole acting responsibly.' Under these circumstances, my dad, having compromised lungs, is staying inside. Being retired allows him to do this.

For thousands of years, parts of the Swedish moose population have travelled the same route between their winter homes and their summer pastures on higher ground. The internal map is not a genetic heirloom; instead, the calves learn about the most efficient paths from their elders, then they remember them from spring to spring. On the show, the experts compare the thoroughfare to a bus route — it gets busier the further down the line you go. For the second year in a row, SVT, Swedish Television, have installed twenty-four-hour cameras around an area where the moose have to cross Ångermannaälven, a river in northern Sweden. In order to keep people entertained at home, when those who are able to begin to stay at home, the network has kicked off the broadcast a week ahead of schedule. We are expected to be bored and in need of moose. The small profile of a moose head against a white square at the bottom-left corner of the screen tells you how many have crossed the river so far, as the streaming progresses, as the days go by and the ice melts, as we listen to Boris Johnson's voice hammer down harder in press conferences and quickly pulverise outside of that screen.

On the first day, the moose-counter says 000, which suggests they must be expecting hundreds. What a cheap way to get people hooked, I think, and bookmark the page.

Slow TV Day 7: There's a recording from last year when one of the reindeers 'photobombed' the Moose Migration. It jumped up and down in front of a camera installed on a tree, its head bobbing in and out of the frame as if on a trampoline. 'Just look at the *eyes*,' I say to Adam. I'm getting addicted to them, glistening conkers on each side of its head. A few years ago, a sound recording of a beluga whale, supposedly imitating a human, fulfilled the same purpose. It sounded drunk and overjoyed, and Adam quickly caught on to how useful it was if I was feeling the opposite, overly sober and under the surface. I reach for this clip with the same abandon, but it's not cheering up that I need this time. I don't want to be entertained.

Quite a few of the reindeer have only one antler. Bizarre, my dad says. I look it up and find that this is perfectly normal. Reindeer are the only deer species where both male and female have antlers, but they don't drop them at the same time of the year; neither do they lose both antlers at once. Here's what John Berger wrote about zoos, photographs, and nature documentaries, the vehicles we use to survey non-human beasts: 'In the accompanying ideology, animals are always the observed. The fact that they can observe us has lost all significance.' We love it when a reindeer 'photobombs' a moose, because that is what *we* would do, given the opportunity. We drag the baggage of humour with us into the image, among many other things.

The WhatsApp group I use to communicate with my dad and sister has become peppered with wild-life alerts. When one of us spots a movement among the trees, the rest of us are immediately informed: 'A reindeer bum!' 'Otter!' 'So many reindeer!' my sister writes. She should really be watching a lecture as part of her physiotherapy degree, most of which has moved online. Occasionally, she needs to go into the hospital to practise on real muscles and joints. Whenever she wears a face mask on the Stockholm tube, everyone around her takes a step away. Because masks aren't widely used, her fellow passengers think she's sick. Oddly, this seems to make the mask twice as effective: protecting the one who wears it, as well as others. As we'll find in a few months, when we return to work in the UK, this isn't the case when masks become the norm. Many will then, unwittingly, regard them as an excuse to relax social distancing. Everyone's individual response is ignited by that of others; we are, and are made of, a chain of reactions.

'Händer det något?' I ask my dad on the phone. 'There were some lovely reindeer earlier,' he says. 'They were all running very fast.' If all you are met with, when opening the screen, is the river's iron skin, or a branch losing snow drip by drip, you can choose to go through the time-stamped highlights underneath the video. Each link tells you about the event in question: 'The reindeer are resting.' 'Otter on block of ice.' Some descriptions are more elaborate: 'Curious reindeer smells the microphone at 11.23.' I'd love to know whose job that is, what kind of tiny power they have, the ones who choose the words to describe what is happening.

Slow TV Day 9: 'Jag har sett järven!' my dad writes. I love it when animals are referred to as *the* bear or *the* wolverine in Swedish, as if all bears were one bear, all wolverines one wolverine, out there and everywhere. I video call my dad and follow his directions on the timeline. There's a rocky slope, blotted unevenly by the sun through a canopy of pines. There are dark crevices, armpits of unknown depth in the cliffside, and a lot of moss. We hold out for the wolverine's arrival. I watch my dad watching the moss, and try to assess his okay-ness on this particular day, how keenly he's feeling his isolation, or if something has managed to sneak in via a door handle (he does take out the rubbish) and into his lungs. A snout becomes visible between two rocks and our eyes widen when a treacle-coloured half-bear, half-badger with a feather-duster for a tail emerges and runs past the camera, licking the ground with its belly, then out of view.

'He's off!' says Adam, who's been watching all of it, and us, from the bedroom door. What do we make of this? It's an English-language phrase I've always liked. As in: what does this become in our hands?

The first time I begin a day by opening the moose-tab on my browser, it already feels like a habit. I head for the livestream first, the 'real time': a time that is real because it is shared by all of us, anyone who's sleeping, eating, struggling to breathe, or having sex right now. The river, my at-risk dad, the nurses, the sick in London hospitals, and the sick in Guayaquil, Ecuador (where I lived aged one to five), where people are currently burying their dead in public parks because there is no room in morgues nor in cemeteries; my family in Colombia under strict quarantine. Then, I scan through the highlights I missed whilst sleeping. This is how, from now on, I clock in these days: first the moose-count, and not till then the death-count.

Slow TV Day 11: The moose finally arrive. It happens on exactly the same day as last year, in spite of the pandemic. The first couple of moose appear on the far side of rows and rows of pines, trudging steadily through wet, borderland snow. There's still snow up north these days, although not much. I can't hear the sound it makes under their hooves, but the look of the snow tells me it's thin and dissolving. We have caught them live. Although no one has caught anything with their hands, the way you catch a fish in a stream or Adam catches moths in our bedroom, the word makes it sound like this is the result of our own efforts, as if there were no TV producers deciding which camera to broadcast from, which view to feed us and when we're allowed to see it.

It's also my birthday. I get a little drunk on whisky, dancing incoherently in our lounge, with Adam playing Rage Against the Machine songs on the electronic drum kit. Following jagged, tipsy lines, I cut myself out from my mid-thirties and paste myself into a much earlier age. Last year, on this night, we were trying to get some sleep on an occupied bridge during a climate protest in London. I remember my legs and dripping nose being a lot closer to the legs and noses of strangers than I would usually have found acceptable, closer than would have been necessary, under normal circumstances, or legal under current ones.

If you're keen to learn more about the moose, there is a recording of last year's 'moose studio' available to watch on the STV website. It was a cosy set-up, with fur-wrapped benches around a fire in a forest clearing, right by the side of Ångermannaälven. Experts were invited to discuss the moose as prehistory, the moose on the eighteenth-century bourgeois dinner table (lung-mash, liver, bone marrow — they prepared some earlier and spend some airtime having small bites), as deadly traffic accident, as suburban neighbour, and in widely shared home videos from someone's backyard. One of the guests is an archaeologist who admits to being called 'the moose woman' by her peers: she explains what the moose meant for the local population during the stone age. 'A moose,' she says with wavy hand gestures which then turn into little power grabs, 'well it's a moose!' It's as if she were talking about parenthood or fire. 'It's a natural force one simply has to relate to,' she says, and I wonder what it's like to share an office with her.

There was a period about four thousand years ago, when the moose almost disappeared from this part of the world, possibly because of a warmer climate. The word 'sagolik', fairytale-like, comes up a lot among the moose panellists. One of the guests is a woman who took to watching a nature documentary about a white moose called Ferdinand during the late stages of her pregnancy. She's a psychologist researching the effects of nature connections on mental health. When she was no longer able to get outside as much as she wanted to, the documentary allowed her to 'flee into that world, and bring that strength' with her.

When this all started, a few elements felt strangely familiar. This is a turn of phrase I've always liked because it should be a contradiction in terms. Most familiar things are also always strange. 'Have you noticed?' Adam said one day. 'We're the party poopers again.' This was mid-March and we'd just cancelled a meet-up with friends at a pub. We said it might not be a good idea because of Covid, and the response was that we could all bring hand sanitiser. 'We're young,' someone else said. 'Even if we get it, we'll probably be fine.' For a while now, we've been the people who bring down the mood at a table, our trying to *do something* coming down as a scoop of unwanted terror straight into people's pints. Adam never brings it up, but when we're asked what we're up to these days, what am I supposed to answer? The bit I find the hardest to explain is that it's not ultimately about me getting sick, just like it's not about me, personally, dying in a flood or fire next year, or in five. The best explanation I have for why I started wearing a face mask before it was mandatory lies in something my dad said to me when I told him I'd been furloughed. 'I feel really guilty,' I said. 'I'm getting paid for *not* working.' 'Why?' he said. 'It's just like being signed off sick.' 'But I'm not sick!' I said.

'The world is,' he said, with uncharacteristic lyricism. I was grateful to him for saying that. It made me remember the points of contact, where we become other, where I become a system.

So, I do find it confusing that the figures in Sweden, a country where civic duty and societal responsibility are supposed to be rooted in our culture, keep rising.

Slow TV Day 12: That woman in the studio loved her white moose, Ferdinand, so much that she ended up naming her daughter after a character in a moose-related fairy tale: *Sagan om älgtjuren Skutt och lilla prinsessan Tuvstarr*, The Tale of Leap the Elk and Little Princess Cottongrass. Not long ago, a collection of Swedish fairy tales was sent to the bookshop where I work. I took it home because I recognised John Bauer's illustrations on the cover. All I know about John Bauer is that he drowned with his family on their way to a new home in Stockholm in 1918. The reason they chose to go by boat and not by train was that there had been a railway accident recently along that route. As a child, I found his images eerie, all the trolls, the tomtar, the golden-haired children, some of them naked and some of them covered from top to toe by their golden hair (I dreamed of being covered top to toe in my hair, which could never be golden — I could never be part of those fairy tales), masses of hair covering the bodies of the trolls too. All these creatures seemed to emerge from a less safe place, just past the sofa or an arm's length from my grandparents' window. The moment the burgeoning nose of a troll began to make the night feel dark, I could shut the book and run downstairs.

Looking through those images now, they are a relief. In John Bauer's work, the wood is safe and intact, as impenetrable as ever. It has stayed the same in spite of everything — in spite of me. Regardless of what I do, how careful I am or how little I'm able to change the situation, this wood is still there.

Slow TV Day 19: Now, we're watching out for the swimmers. The moose-counter in the corner — the one that keeps track of how many have crossed the river — remains on 000. At about half past nine in the evening, three moose wander down a steep slope at Södra Udden, and one of them takes the lead all the way to the water's edge. The night is bright in the way only nights up north are lit — un-dark — with the sun sifted out but leaving light's basic elements. All three moose, two of them possibly calves, are now up to their haunches in the water. They stay there, taking sideways steps and little shimmies, sipping from the lake. After a short while, the bigger one turns back toward the shore and starts walking up the slope again, with the other two quickly in tow. The counter doesn't change from 000. 'Fegisar,' my dad says. Little cowards.

What do nature documentaries really document? The work of David Attenborough, whose renown also reached someone who grew up in Sweden and Colombia, has faced criticism over the years, on the grounds of including 'staged scenes' (a polar bear giving birth in a zoo as part of *Planet Earth*) as well as portraying a false image of wilderness as untouched, guiding the viewer's gaze away from annihilation caused by human activity (*Dynasties*). What would an entirely authentic nature documentary look like? Essayist Amber A'Lee Frost wonders if it would be 'ethical (or even possible)' to install cameras in a polar bear's wild den: 'Is a polar bear giving birth somehow less "wild" in a zoo?' What kind of authenticity are we after when looking at non-human animals? How could we possibly extract ourselves from the viewing that happens through our eyes?

Relying solely on a series of cameras installed in trees and in the moss, The Great Moose Migration seems to avoid the most obvious pitfalls — it doesn't give the moose human voices; it refrains from adding dramatic music, or staging battles between moose (although these happen — they get nervous when too close to each other). Does this mean that, as humans, we aren't there? The migrating moose, like Bauer's images of children and trolls, calm me, cradle me in my seat. There's so much we don't know about this virus. Watching the moose, these days go by almost quietly, as if they didn't have such insidious corrosion rushing through their veins. 'Diseases have always come out of the woods and wildlife and found their way into human populations,' a *New York Times* article tells me. 'And with modern air travel and a robust market in wildlife trafficking, the potential for a serious outbreak in large population centers is enormous.' That was 2012. But these are moose, not pangolins or bats. Is that why I'm comforted? Is this wild life safe, just like we are supposedly safe from the worst of climate breakdown, for now, in this section of this hemisphere? Does this mean that we mistook the forest for the fairy tale? Or that we're looking for something which is no longer there? Although presenting itself as objective observation, not storytelling, the very premise of the show relies on a narrative of humans peeking into a peaceful, timeless space in which we aren't implicated. If we are not a part of that story, that system of interlinked reactions, we don't need to take responsibility for what happens inside it.

I understand why someone would watch a white moose when trying to calm their uterus, which is about to expulse a new human.

Bauer didn't use much colour in his work. According to the book's introduction, he felt that subdued greens and browns better suited the woods of southern Sweden, which was where his art came from, and where it will always live. It could be that simple, that a part of me is homesick. I don't think it is that simple, or that homesickness ever is.

Slow TV Day 25: By now, six moose have crossed the river. I missed the first few, as it happened during the night, and I'm not yet hooked enough to keep watching past midnight. I can't say if this is a sign of sanity or shows a lack of commitment.

With regards to moose-viewers elsewhere, it takes a while to trace the watch-party. The hashtag comes in so many variations; it too migrates: #denstoraalgvandringen (for those without Swedish vowels on their keyboards), #denstoraälgvandringen, #moosemigration, or simply #moosewalk, which sounds like it was born in the eighties. Someone called Papa Emeritus tweets a screenshot of the river along with the hashtag #skattepengar and #slöseri, #taxmoney and #waste, but mostly there is gratitude. 'Patience is rewarded,' someone called Elisabet writes. '#SlowTV is great #EntertainmentAtHome while battling Covid-19 #SVT'. A viewer in Brazil shares that sentiment, saying on the last day of the broadcast (according to Google Translate) that from now on 'quarantine will be sad'. The programme also seems to attract viewers from opposing sides of the political spectrum. Göran Bengtsson thinks the Moose Migration is the only watchable programme on SVT nowadays, amidst all the 'leftist propaganda', whilst Non Servi Ⓐ m ANTIFA, an account full of anti-Trump tweets, writes: '10 points to #SVT!' They're both, apparently, finding something to love.

Slow TV Day 26: Moose are great swimmers and they're usually fine walking on ice, but the in-between stage, ice which is about to melt, is a problem for their skinny legs and sharp hooves. At 11.08, a small group of moose begin to cross the river, which is still frozen at the edges. Three of them are evenly distanced in a row, with a pair holding the rear. As the first two step onto the opposite shore, number four makes an unexpected, sharp turn, and continues to swim along the beach without getting any closer to it. When it finally decides to head for land, it has reached an area with a broad streak of ice lining the beach. It begins to swim straight into that ice sheet, hoisting itself up on its front legs only to pierce the frozen layers and crash through them each time. It tries, again, to get up on the ice and collapses, heaving, then falling. It finally stops and just lies there, half-in and half-out of the water, panting.

What do I see? That it looks like serious pain, the kind you cannot distract yourself from. What do I want to be told? That this would have happened regardless — that because I am here and the moose is there, that because it's a moose, what I do has no bearing on what happens next.

Göran Ericsson is Professor in Wild Life Ecology at the Swedish University of Agricultural Sciences. He was a regular participant in the moose studio in 2019, and has since set up an online question-box about moose on the university website: 'What are the causes of the reduced weight of calves since 1985?' someone has asked. 'My daughter wonders if the moose have cold feet.' 'Can moose get corona?' I find it welcoming, that someone who incidentally watches moose with their breakfast gets to ask their question alongside those who are well-versed in breeding patterns. In an email to Göran

Ericsson, I ask if he's noticed a difference in the kinds of questions people submit this year. He tells me that, this time, the questions are more fundamentally about wild animals, more generally about nature. He's seen a definite rise in interest, he says. His gut feeling is that the programme is providing an entrance to 'something calm, comforting and perhaps, lagom boring'.

Oh, that word. Lagom. It's one I've been thinking about too, during these weeks of moose watching, news watching, watching oneself and others from afar. The word lagom has a reputation as being non-translatable. It means not too little and not too much: a lagom portion of meatballs, being lagom warm. Just right. Swedes sometimes refer to our own country as the lagom country: supposedly, we don't like extremes. During both world wars, Sweden remained officially neutral. Closing primary schools, the state epidemiologist says during a talk show, will likely lead to worse peaks when they open again. So, you wait and you don't close the schools. You follow common sense and you don't make sharp turns. Lagom is being not hungry but not too full either; it's the greyish green and uncertain brown of John Bauer's woodland images. That neutrality, though, is only one narrative among many, and however much of a lone ranger it likes to think it is, it is wired into all these other stories. German transports of war materiel and soldiers to occupied Norway were allowed through the country during WWII. In the seventeenth century, Sweden was a European superpower, waging bloody wars of expansion. The north of Sweden, the setting for our moose show, is the stage of historical state oppression of the Sami people. Sweden was also the first country in the world to establish a National Institute for Race Biology in 1922. These are colonised lands, rarely regarded as such, and there's nothing lagom about any of it.

As for the moose, they're built for cold. At higher temperatures, their energy is rerouted to regulating body heat, and the food itself becomes less nourishing. There's been a reduction of 20–25 per cent in moose reproduction in Sweden over the last twenty years. They're also more susceptible to certain diseases. In 2016, it was confirmed that so-called 'brain worm', a parasite which also infects reindeer, was spread throughout the country, and is likely to become more common with rising temperatures. It attacks the moose's central nervous system, causing them to become disoriented, wobbly, lost, and confused.

Slow TV Day 26, a bit later: The moose makes it safely onto land. It finally reaches thick-enough ice to hold its weight and lugs itself up on its folded front legs, then the rest follows. There's an inaudible sigh of relief on Twitter.

In 2018, Attenborough said that his documentary *Dynasties* would be 'a great relief from the political landscape which otherwise dominates our thoughts'. *Den stora älgvandringen*'s relationship with its viewers, however, suggests that rather than a relief from politics, a balm to the frazzled, it offers a continuation to our preferred narratives, and politics. In the UK, a 'community spirit', a 'pulling together' (our WhatsApp group full of neighbours we've never spoken to, clapping for the NHS, the comparison to a 'Blitz Spirit'), will all become stories told about this crisis. They in turn exist within a wider narrative, in which it's easier to clap for the NHS than to fund it. It's a story which anyone existing within late capitalism inhabits, including (and this bears repeating) Swedes: a story in which individual freedoms reign supreme, and individual responsibility is defined and curtailed by these. In spite of all the talk of trust in authority, of Sweden having access to this unique cultural common sense, that too is only one of several stories determining our response, how we react when frightened, and the shapes power takes. Which of all Sweden's narratives explains why the country has one of the highest per capita fatalities in Europe by the end of the spring? It sounds obvious, but Swedes act selfishly too, in an individualist system. It's only when it doesn't fit into the narrative we expect, and desire, that we call it a human, instead of a Swedish, thing.

How natural it is, to carry it all in with us: the anxiety, the fairy tales, the need for reassurance. Be it because I am, at least in part, Swedish, or because I'm a person in real time, right now, I desperately want a normality which shelters me, and the people I love, from the extremes I know are there. I want trolls in my woods but no starvation and no racism. I know this is what I'm looking for, and I know it doesn't stop there.

'The image of a wild animal becomes the starting point of a daydream,' John Berger wrote, 'a point from which the day-dreamer departs with his back turned.'

Slow TV Day 30: Someone has described today's highlight thus: 'Moose hides behind a tree, 13.13.' My dad thinks the moose looks absurd, the way it stands so still behind the trunk of a tree whilst facing the camera, or the direction where we know the camera is placed. All we see of the moose is its belly puffing out on one side of the trunk and an ear on the other. If it is indeed hiding, it's doing a poor job. Adam comes up behind me and begins to narrate the thoughts of the moose in the voice of a confused person who, I imagine, wears a top hat but no pants. 'You can't see me,' our cartoon moose says. 'I'm invisible!'

Turning our backs on the daydream, where do the day-dreamers *go*?

When it happened: when my mother's nerves ended and felt nothing ever after:
I felt nothing, until we were already stuck in the after.

The feelers in both our spines were asleep, in bed as we began but under different ceilings. And it bothers me, that my nerves didn't stretch far enough, weren't open enough for this,

when those in her back would feel me dancing. I sent a text about nine, saying how I hadn't liked leaving her in the hospital room. We'd be back first thing. No one (they said) could have known what was coming. 'She'll be asleep now,' my dad said, and switched channels. We turned the light off, with that signal

of an unread message, grey and leaking hers

but not our ending.

If there are times when I can feel them

If there are endings carrying still

What is to say there is no end

To what can possibly be felt?

A NATURALISATION

1.

I was recently made natural. I needed to be dragged through something viscous with the mud of bureaucracy, the artifice of national acceptance, in order to become a part of this place called the UK. My citizenship ceremony took place shortly before the national lockdown in March 2020, which was why the chair of the council, who conducted it, was wearing black gloves. Possibly, they were velvet. She explained that she would not be shaking people's hands as she normally would. There was a picture of the queen, blown up but unpixellated, mounted on an elegant wooden desk at the front of the room. The prospective citizens were escorted to our chairs, whereas our guests (two per new citizen) were shown to the side benches. It was explained that, although we couldn't shake hands with the chair of the council, we were still welcome to have our picture taken as she handed over the certificate.

I hadn't known this was a thing people did. I didn't think I had particular views on it.

After the ceremony, Adam and I went to get some cake and took it back home. I was no longer an inserted element, no longer unnaturally home, but home truly, according to this certificate I had in my hand, with a message from the Home Secretary. The whole process, from unnatural (artificial?) to natural, had been finalised in a room with heritage furniture. The room assaulted anyone who entered with a historical aroma, which must be part of its naturalness, rooted in deep, inscrutable pasts, the kind you cannot know without a degree. The furniture was natural as certain squirrels are natural, meaning mostly not at all, as the majority of British squirrels are not native to these islands.

Back home, we celebrated by drinking coffee from my Colombian coffee cups, the same design my parents had when I was growing up. It was as if to say: this is far from over.

2.

So many of my British-born friends are angry with their country. Them being my friends and them being angry with their country's government, and half of their country's population, are linked conditions. It says something, that I don't have friends who are happy with their government at present — something about who speaks to whom, and what they allow themselves to speak about in each other's presence. It's as if sometimes the country itself reacts with raw flesh inside the mouth and the name comes out smoking. There's been a collision in there, a breakdown. Not only has the government become synonymous with the country, but both are repelled by the self.

Often, when I told them that I was applying to become a British citizen, they would roll their eyes. It was never at me, but at the process and its reputation; some would laugh sympathetically and tell me about someone else they knew who'd gone through it, some complication there had been, and how, often, it was still ongoing. Some asked me why on earth I would want to become a citizen of this place which is 'going to hell in a handcart', and some joked about me heading in the wrong direction. 'Aren't people wanting out?'

They were applying for Irish or German passports based on family connections. They were attempting to be made natural elsewhere, rejecting a nation which, in turn, had rejected some of their most fundamental beliefs.

Over the course of a few months, I carried around a study guide, in preparation for my Life in the UK test, the passing of which is one of the requirements for being naturalised. The study guide includes a very short history of the UK, from the stone age to Brexit, as well as lists of famous artists, athletes, and musicians, and explanations of the UK's legal system and devolved parliaments. People around me would ask to be tested on its contents. When questioned on the events leading up to the Glorious Revolution of 1688, or the difference between Crown Courts and Magistrates' Courts, they said: 'no British person knows this.' The inferred meaning, I think, was: 'we're sorry that you are required to know this in order to become British, when we are not.'

Supposedly, there's no more natural way of being from a country than to have been born in it. If most people born in the UK aren't able to answer the questions included in the Life in the UK book, a condition for naturalisation, what did this natural state, which I must be longing for, consist of? How could it be defined? What was I asking for?

3.

At the time of my ceremony, face masks hadn't yet become common. There were one or two people wearing them. When the Chair of the Council welcomed us, she made a point of congratulating us for having shown up, in spite of the situation. By being there, she said, at a time like this, we were 'showing true British spirit'.

'True British stupidity,' Adam commented, afterward. He was born and raised in Yorkshire. His grandparents came over from Lithuania during the war, but his grandmother may have been Bulgarian. We don't really know.

Sometime later, I think about how quickly that changed: what constitutes a 'British spirit', with regards to attending public gatherings.

4.

Recently, I have been thinking of transplants. I've been waking up, folded between two kinds of restlessness, one sweaty, one dry, and thinking of how the new organ, if the timing is right and the doctor good, is embraced by its new surroundings, moving from crisis to survival. I've been picturing thousands of tissue strings and electric connections courting the new family member and saying:

'Yes. You can stay!'

What makes the surviving body, changed like this, less natural than any other? What does it take to be accepted as whole?

5.

It's true, however, that I didn't, technically, need to become a citizen of the UK to be allowed to stay, at least not at the time. After the Brexit vote, I was granted 'settled status' by virtue of being an EU citizen and having lived here over five years. I'm also married to a British citizen, which helps sometimes, though it's far from always enough, especially if you're a person of colour, especially if you have a low income or no income at all. For these reasons, there are countless layers of anxiety, horror, ignorance, and cruelty between my situation and that of those who risk deportation. I had to provide documents, but they were documents I could get hold of. It was a faff, but not impossible. There was some waiting but for months, not years. There was no intimidation and no threats, no abuse. The steps I was asked to take in order to become a 'natural' inhabitant of this chunk of land pale in comparison to the hurdles thrown in front of others, who come from places the government doesn't want you to come from. These, in turn, are little compared to what they do to those they deem 'irregular', who aren't supposed to be part of the same system as the rest of us. They are so unnatural that there's not supposed to be a way in for them.

Does the fact that it was easier for me, for whatever reason, make me more natural to begin with? Does this suggest that I should leave or that I should stay?

When they said that I was going in the wrong direction, I thought (and didn't always say): because I can. The day after the December 2019 general election, in which I had not been able to vote, I had dinner with some friends. They said they felt done with it. Their anger with the government, and the people who voted for it, had resulted in a rejection of nation as geography.

'What about everyone who can't leave?' I remember saying to them.

I was trying to say that leaving because you disagree with your government is far from the same as being forced to leave. I meant that the people who suffer the most from austerity and isolationism are also always the poor and vulnerable — who can't pack up and try pastures new. There seems, to me, to be a logical gap between knowing that you can and assuming that you should (leave) — or acknowledging that you don't have to and assuming that you shouldn't (stay).

Having said all this, I also left because I could, and because there were things in Sweden that I didn't like. I also conflate disagreeable parts of that country — the things that always made me feel, as my mother used to say, like un mosco en leche — with the country itself, that elongated piece of land.

6.

On the chair, to my left in the Guildhall chamber, was a man, possibly in his late forties or early fifties. I suspected he was smiling before looking at him, then noticed that he wasn't smiling at all, or that the smile, if it existed, wasn't on his face. The excitement lived in his posture: the feeling that if I touched his jacket, I'd be jolted by silent, static electricity. Shortly after they mentioned the rules about handshaking and photographs, this man leaned a little bit closer to me. He held up his phone half-way between my left and his right knee, and asked in a whispering tone if I'd take a photo of him when it was his turn to go up there. He looked ebullient, in a shy way, a jitteriness around the middle reaching out to me, perhaps expecting me to be just as excited and to meet him there, both of us celebrating the end of a chapter.

I said that of course I would. He then asked if he could do the same for me, upon which I motioned toward the far side of the chamber, where the top of Adam's head was only just visible behind a row of other people's relatives. My Person appeared to be looking intensely at the image of the queen. This time the man did smile. I would have liked to ask him how long he'd been waiting for this, to offer something equally tender, and equally simple, back across the gap between our chairs — to know if this, indeed, was a very good day.

7.

When you're granted citizenship, you are sent a letter inviting you to book a time for your ceremony. When I called to do this, from the basement at work, they told me that the next date available was the eighteenth of March. My mother died on the night between the eighteenth and the nineteenth of March. This would make the day of my naturalisation the second anniversary of my mother's death. Casually, like this, I wasn't sure what to think. 'Does that work for you or would you like to wait till next month?' the council person asked, and I replied that I'd like to book it. I discussed it with my sister and my dad over the phone, who both agreed that my mother wouldn't have minded.

My mother went through her own naturalisation process, although in Sweden they don't call it naturalisation. For a long time, she didn't have Swedish citizenship, because Swedish law didn't allow dual passports back then. Becoming a Swedish citizen would have meant giving up her Colombian passport, which she wasn't willing to do. I can't remember how she explained her stance. She might have said something about having a bit of yourself chopped off. Once the Swedish government changed the law, my mother got her Swedish passport, having lived in the country for twenty years. She wasn't required to attend a ceremony; we didn't have a picture of her next to a flag. More than anything, it was a practicality, because travelling with a Swedish passport was, and remains, much easier than travelling with a Colombian one. Customs officers are simply a little less racist.

These small booklets, with their thick, conspicuous pages, reveal themselves repeatedly as inhumane tools, controlling the most human experiences of moving and of being moved. To speak to my mother about my naturalisation may have gifted us something else: a way of speaking about her experience of being a young brown woman in Stockholm in the 1980s. I remember stories of racist microaggressions, of the people staring in queues and the family acquaintance who looked at my dad and said, 'does she speak Swedish?' She talked about her fear of skinheads. I remember how, in 1991 during the months when John Ausonius (known as Lasermannen due to his use of a rifle equipped with a laser sight) shot eleven people, mainly immigrants, she wore a dark coat to be able to spot the red dot. As a child I didn't have a historical scaffolding, a structure of sense-making in which to place any of this — they were solely stories of evil, simpler for that. Speaking to my mother about our citizenships now would have meant building a new world together.

I haven't had a Colombian passport since I was a child. When I asked, as a teenager, if I should renew it, my parents said that when visiting Colombia, a Swedish passport would be safer than a Colombian one. I didn't question this — like nationalities, assumptions about one's belonging, the attitude evolved around a home, can also be inherited.

This naturalisation business was always a question of practicality (more on that later), but never just that either. It was never easy and never something that grew, like a very innocent poppy, out of a grassy hillside.

8.

I never knew this, but when a heart is transplanted from one body to the other, it loses all nerve connections, and remains thus, sometimes for many years. Of course. Yet I always assumed that the nerves grew back, and had to do so in order for any heart to function. I'm surprised when an article in *European Heart Journal* tells me that the 'majority of patients remain completely denervated during the first 6–12 months following transplantation'. I question the validity of my source when I read in a 2002 study featured in *ScienceDaily* that reinnervation, the regrowth of nerves, 'did not significantly impact survival'. I spend fifteen minutes trying to *feel* my own heart.

There is a fundamental comfort in this, that body recognises body — that the relationship between a part and the whole is far more complex than any individual bridges we can point to and say: that's where it happens. It's basic, yet monumental, that we can adapt this way.

9.

The Home Office was formed in 1782. Before then, two secretaries of state dealt with foreign issues divided geographically: the Southern and Northern secretaries. The Home Office was created in order to focus on domestic matters. It deals with the 'inside', but it's not all inside matters it holds under its wings; it doesn't respond to broken bones or domestic abuse, and it doesn't concern itself with what we eat or how we educate our children, or put into their bellies, the most internal of places. Other than migration, it's also responsible for tackling organised crime. The pairing of these two things makes it, conceptually, more like a house-alarm system than a house meeting, called to decide whose turn it is to do the dishes. Its remit is only that aspect of the home which can be breached.

'A man's house is his castle,' Adam declares, sadly, when I run this past him.

He has a history degree and, he says, that's the only reason he remembers even a fraction of the stuff in the Life in the UK study guide.

As far as I can tell, the familiarity between crime and migration politics is far from exclusive to the UK. France's Ministère de l'Intérieur tackles organised crime as well as migration. The US has its Citizenship and Immigration Services, part of the Department of Homeland Security, which also oversees the counteracting of terrorism; it was formed by George W. Bush after September 11, 2001. In Sweden, the police, although a separate unit, belongs to the Ministry of Justice, along with the Migration Agency. In *The World of Yesterday*, Stefan Zweig described the contrast between his pre- and post-WWI travel experiences. Whereas before 1914 he'd been able to travel 'to India and America without a passport', after the war, 'all the humiliations previously devised solely for criminals were now inflicted on every traveller before and during a journey.' What would he have said about the fact that in Sweden, it's the police that issues passports? In Colombia, on the other hand, migration is the responsibility of the Ministry of Foreign Affairs, as is diplomacy. What an outlier, in my small and highly subjective survey — connecting migration with being polite to foreign powers, instead of assuming foreigners will make the home less secure. I need to look into other countries with fifty years of civil war in their baggage, to find out how rare an example Colombia is, especially in the last few years, with the influx of Venezuelan refugees. How much diplomacy is awarded to them?

By using the word 'home', a government entity colonises it, deciding who it encompasses. It manipulates the already permeable space between the ecological and the bureaucratic, the organic and the artificial, moulding the relationship to its own agenda. It defines the very idea of 'home' as exclusive. Its policies separate those who can assume to sleep, cry, live here on the basis of being alive, from those who have to book an appointment first.

The name itself seems to say: if you belong here, you will be safe. We are the ones who decide if you do.

10.

Next, there was a speech, delivered by the chair of the council. It included an astonishingly quick run-through of British history, during which the union of England and Scotland in 1707 was described as the nations 'coming together to form the UK', before moving on to list some of the country's most famous writers. 'Came together,' was Adam's reaction to this, on our way out through the Guildhall corridor's parade of portraits. 'The signatories were chased by an angry mob down the streets of Edinburgh,' he clarified. The speech reminded me of a passage about the Hundred Years War, in the Life in the UK study book. 'In 1453 the English left France,' the text said. The English, by the sound of it, were a little tired and didn't care to stay any longer at a party. *When did the English leave? They left around midnight.* The mess of it is smoothed out so easily among heritage furniture.

The two countries 'came together'. The 'English left'. Language lays a soothing film over the burns of history and makes violence seem less, as it makes what's artificial appear natural. What passes as natural becomes unquestionable, such as the 'home' in Home Office. It's free to hide reality as it's being lived — the atrocities of imperialism and everyday oppression as well as all the ties we really do hold.

11.

Before you attend your citizenship ceremony, they ask you if you'd like to either swear an oath or affirm your allegiance, the difference being that if you choose the former, you promise things 'by Almighty

God'; with the latter, you 'solemnly and truly' declare them. Not being on talking terms with any Abrahamic god, I went with the latter.

The declaration was printed on my letter, so that I could prepare for what I would be promising. Roughly half of the congregation read the oath, some muffled by their face masks (what if they didn't read it at all? What if it was like me as a child at mass in Colombia, head down, mouth defiantly shut? I'd not thought of this before, but would face masks offer a loophole, a 'yes but I had my fingers crossed behind my back' kind of excuse?), others looking intently at the queen's portrait, or at the ceiling. Some of the relatives leaned over in their seats, sniffing for the words themselves, perhaps for a change in body odour from one moment to the next. After the oath and affirmation, there was the pledge, which is read by everyone, and which makes you, officially, a British citizen.

This is what I pledged:

'I will give my loyalty to the United Kingdom and respect its rights and freedoms. I will uphold its democratic values. I will observe its laws faithfully and fulfil my duties and obligations as a British citizen.'

Having affirmed this, having pledged thus, having paid £1,400 pounds plus the cost of travel to Cardiff to have my biometric data checked, having managed to contact old employers who agreed to testify that I was here, three or five or seven years ago, that they saw me and I was who I say I am, I was naturalised. My feet were ready to sink into the earth so that roots could take hold, like British dirt's business was my business. We all sang the national anthem.

12.

It's not so much the reality of a transplant — how it feels — that intrigues me, and I feel bad about this. I know people who've had heart surgery, but they still carry their old heart. My own memories of hospitals are too entangled with love, guilt, and confusion, with trying to care for someone and caring in the right way, to allow this to become a medical interest, one of detailed rights and wrongs, dos and don'ts. So, I make it symbolic. I'm drawn to the notion of one organ becoming an inextricable part of a new body, and how we can think about other kinds of transplant through the lens of that truth: that it works, and we are neither fixed nor impermeable.

Obviously, transplants — of self or of parts of the self — are never abstract, never only symbolic.

13.

One by one, people got up from their seats, approached the desk, then sat down for a few seconds as their names were verified by a clerk. One woman, possibly around my age, had a small child with her. They were both being made citizens at the same time. When I think of them now, I see them dunked into a pool of water, hauled up changed, even though nothing of the sort took place. It's as if all this has made me go biblical.

There was someone who had to take off their mask so the clerk could hear them. When the man next to me got up and received his certificate, I took a picture of him standing next to the chair of the council, with the envelope in his hand, then I took another one as he was receding, just to be on the safe side. Hopefully, at least one turned out okay. It was midday and I declared, to myself alone, that above all I was hungry. It seemed ludicrous that people would ever get hungry in a room like this — hunger, like pain, wasn't part of the process.

I walked to the front and reached out for my document.

'Thank you,' I said, then I turned back toward my seat, quickly glancing at Adam, who looked neither proud nor appalled. He's not the reason I'm here, but he's part of the reason I'm here. 'No picture?' said the chair, her head tilted. She smiled and looked out at rows of seats, in search of a phone, belonging to me and mine.

'Nah, I'm good,' I said, and shuffled back to my seat.

'Your dad might have wanted one,' Adam said afterward. I reminded him that my dad is very practical, and that during a pandemic, standing close to someone only for the purpose of a photo seemed highly unpractical. But of course, others did stay up there for the picture. I know nothing about them. I'm oblivious to how much they had to do, how long they were made to wait, which sections of themselves they were asked to carve out, give up, make obsolete, and say goodbye to in order to fit the mould of the naturally British. I know nothing about the pain or joy that brought them here, and I don't know if they had a choice.

Which is to say, not that I should have stayed for the picture, but that the picture, the uncomfortable furniture (which by the way reminded me of a Swedish saying: att ha träsmak i rumpan, to 'have wood-taste' in your arse, when you've been sitting too long), the songs, and the ruffles are things we are made to want. It's part of the artifice of belonging, that we are made to desire them. I didn't want to have to desire them.

14.

'The body is not national,' writes Danish poet Pia Tafdrup.

She was an important poet to me, in the specific, stripped way that only poets discovered at the end of your teens are important, neither more nor less impactful, but never the same as later in life.

I'm reading her in English now, which is not unnatural, but new. 'The body is not national'.

I happen to read this poem around the time of a protest outside the Home Office. We're there to highlight the connection between climate collapse and migrant rights. Some of the speakers are currently going through the immigration system, several as asylum seekers. One of them calls in from a detention centre. A mobile phone is held up to the microphone and we hear their voice from inside a far-away 'somewhere', the least homely place, organised and managed by the people who take it upon themselves to dictate our relationship, as citizens, to 'home'. This stranger's voice bounces off non-place walls, invisible to us on the street outside the Home Office. He speaks from a room we're not supposed to look inside, into which people are disappeared because this spectacle called 'home' doesn't include them. Soon, the police begin to move through the crowd.

'The body is not national'.

I may have forgotten that poetry is about conjuring a reality the way you'd like it to be. I may have learned that the way you'd like it to be sometimes obscures the lucky circumstance that makes dreaming possible.

15.

When they asked why I applied for that piece of paper, I gave a few different answers. Sometimes, I swapped between them; most often, I offered all four:

I chose to come here meant that when I arrived in Scotland, I found people who made me feel chosen. I might have meant that there is love involved, but like love between people, love of a place is relational. It happens between you and a place, because of how it's treated you, not as adoration of a calcified object.

Because in this climate, you never know referred to the political climate, as well as ecosystems collapsing, both of which may lead to my situation in this country becoming unstable. They, the ones who decide if this is my home or not, are prone to changing their minds.

So that I can vote.

So that I can get arrested. That one defused whatever sombre vibe had settled over the conversation. By this I meant that being allowed to stay is not the same as being allowed to live. It sounds both simple and ludicrous, it made people laugh, but when state powers are threatened by dissent, some people are picked out in a crowd because we're already low-hanging fruit. Whereas a white person, or someone with no foreign accent and a British passport, someone who isn't queer or visibly disabled, is seen as a nuisance when blocking the entrance to an oil-company headquarters (or staging a protest outside the Home Office), anyone who falls short of a state ideology's version of what is 'natural' — which, really, becomes another word for 'desirable' — quickly loses their right to call Britain home.

We're allowed to be in our home at their discretion, but not allowed to care enough about our home to improve it. 'Home', as defined by the Home Office, being an idea that does not survive well when it comes to a global crisis.

16.

I did a social-media poll while I was writing this: 'Word play (humour me),' I wrote. 'What would the Home Office be called if we kept the word "office" (they do have an office) but replaced the word "home" with something more descriptive of what it does? Any brain-waves welcome.'

The responses, from within my bubble, included: the Heartless Bastards Office, the Palaver Office, the Office of Coercive Control, the Home Isn't Here Office, the Not Your Home Office, the Divide and Rule Office, the Hostility Office, the Theresa-May-doesn't-like-foreigners-and-worked-here-too-long Office, the 'We don't like 99% of the people in this country' Office, the Alienation Office, the 'Go back home and stay the fuck out of our office'.

The humour was cathartic. We can make fun of the rituals and the pomp. Perhaps some do because they don't wish to look at what happens underneath, what is hidden behind the circus — others because they are so angry at what the rituals represent. My friend Pete suggested the Alienation Office. I wrote back to him saying I thought it was the most accurate reply I'd received: the Home Office's veritable purpose is to create groups who are defined as aliens, whose bodies are an addition to the place, not part of it. 'Pretty much, yeah,' Pete replied. 'But I think there's also a double action in that it alienates the country from the rest of the world too, from any sense of global responsibility.'

Global crises — the diseases, the food insecurity, the extreme weather and political collapse — have the ability to stretch out ideas of 'home' beyond themselves, pulling them toward opposite ends. Facing a crisis which demands a global response — not as in eight or so elite governments telling the rest of the world what to do, but as in the only world we know, responding as the body it is — 'home' can shrink or it can expand. It can turn into its own dark nights, close the doors, and set the alarms, or it can stretch so far there's no imagining where it ends. Borders can be used as shields against uncertainty or be recognised as the artifice they've always been. What's natural is the mobility of life, the migration of us and of parts of us.

17.

We felt this should be celebrated, although we weren't exactly sure why. It did have something of the adult baptism about it. I had half-expected people to put their hands on my head and say, 'you're a woman now', if we weren't in the middle of a pandemic, if it weren't for that visible crisis. As a teenager, such pronouncements would have made me feel equally uncomfortable.

If nothing else, perhaps my spending £1,400 on it, the hours collating payslips, and the tracking down of old travel dates made the end of it worth a little feast. They told us there would be a 'small gift' in our envelopes, but curiously there was only the certificate, and the letter from the Home Secretary.

'Did you get a gift?' a short person asked us in the hallway, and the three of us chuckled.

Adam and I went to get two slices of cake and brought them home. In the evening, we shared a dinner with my dad, my sister, and her partner over Zoom. It was at the very beginning of the pandemic. There were congratulations, a toast for my mother, her picture on the table, on the anniversary of her death, and no one brought up the fact that I was no longer, officially, from the same country as her, that some contact had been severed, without which we still belonged to each other.

Postscript, December 2021

Three days ago, the UK government's proposed Nationality and Borders Bill passed its third reading in parliament. The bill will further criminalise the ways in which humans have always, naturally, moved for greater safety and wellbeing. By penalising the manner in which people reach this country, it goes against the UN refugee convention of 1951. A recent amendment also facilitates stripping citizens of their British nationality without the need to inform them, as long as they are 'eligible' for citizenship elsewhere, a move which will disproportionately put people of colour and other minorities at risk. The last year has also seen the progress of the new Police, Crime, Sentencing and Courts Bill, currently in the House of Lords. It criminalises GRT (Gypsy, Roma, and Traveller) communities' way of life, and increases stop-and-search powers. Its latest amendments include the creation of a new protest-banning order for those with 'a history of causing serious disruptions at protests'; broader, more flexible definitions of what 'serious disruption' means; and the introduction of 'interfering with nationally significant infrastructure' as an offence, punishable by twelve months in prison, a fine, or both. In this nation, as it is officially and intentionally defined, few examples of infrastructure are more significant than the Home Office, a detention centre, or an oil terminal. These are also inseparable from the climate crisis, and global climate injustice.

These laws, movements in the direction of authoritarianism and protection of capital, were never separate, and they ride along an international current, felt in different ways in all my homes. They are tools of a racist, xenophobic state which, amidst intersecting crises, invests in confusion and fear, setting out not only to promote an oxygen-deprived version of home, but to deliberately disconnect us from our greatest, most vital responsibilities — those that tell us we are here to protect all of it.

Why   so nervous?

How could you not be this intoned, this much like a song, when you're knitted with them: the smallest combustible meeting places are in your skin. It's like saying you aren't porous, or that bread and sweat are maladies.

If what you are is a pack of them: ett nervknippe that doesn't end at its borders, that reaches for the world nervously, which is to say, by way of these and other endings.

My Person comes in wearing a tree on his shirt (the nerves); there's a picture of a friend's hand (the nerves); a shot of the Mozambique floods from above (a cameo of the nerves); a petal's blood work and the purposeful veins of light. When I thought that I, like my mother, might have a tumour in my head, the optician fired air into my eyeball, then offered it up on a screen for us both to marvel at: an estuary of tangled yarn.

('Are those nerves?' I said. The optician gave me a look
as if to say, what do you know about the nerves?)

It's shameful, a friend of mine said, that you're allowed to tell someone:

'you feel too much'.

MIXED SIGNALS: FIVE MOMENTS OF UN-BELONGING

1.

One second, it's possible for us to unfold; the next, that possibility is gone. I am at work, and at work I have a phone voice which is different from my normal voice; it's even different from the voice I normally use at work. A warm thought, here, to anyone working in a call centre. I wasn't working in a call centre but a bookshop. My colleagues spotted the phone voice when I'd just started and gave it a name: Jessicaaaah, it's called. Jessicaaaah is her own template, complete with the long 'aaah' trailing off like a cape down the staircase my Swedish grandmother dreamed of, if ever one of her grandchildren married an English lord from one of those TV shows. Jessicaaaah is bubbly (why that word? People are not sodas) but also confident enough not to leave room for questions between any of her standard moves, no spaces in which to wonder about the gaps:

'Is there anything else I can help with?' she says, but this time the answer is too quick.

'You're not in your homeland,' says the lady.

Her voice is polite, so presentable, that my hand goes straight up to my nose and wipes it, even though we're on the phone. All I know about her is this voice, elevated to the rim of a fancy glass, and the five books she wants to send separately to different friends around the country, a few 'in Europe'. Sometimes, the phone line goes into spasms, but we were getting along just fine. Everything was steady, clean, and intensely regular.

'Well,' I say and laugh into her ear. 'I'm a British citizen now, so I guess I am.'

That's Jessicaaaah, going with the flow until everything is nice again, until we are once again safe. There are five miniature staples in a mystifying pattern on the desk. Someone has been making shapes whilst on the phone, twirling metal between fingers (whose fingers? I'm imagining all my colleagues' fingers), searching for the most efficient answer, perhaps sticking one end into a thumb. Whichever one of my colleagues made this mess, they were not asked if they were in their homeland. Jessicaaaah is not supposed to be asked. She laughs. I laugh.

'Oh, I just meant originally,' says the lady. 'I'm very interested in accents, you see.'

If it weren't for the ovulation pain, which kindles the same side of my body as her voice on the phone, or for how the question wasn't a question, but a demand for confirmation of what she already thinks she knows (not 'are you not in your homeland?' but 'you're not in your homeland'); if it didn't make me think of the HBO series, I would have asked her what she meant. Home-land as opposed to any other kind of land, as opposed to another's homeland. You are not in *your* homeland but in mine. I could have asked.

'------'

If not for our home, Adam's and mine, and that it is on this land; if it weren't for the sixty-four environmental defenders killed in Colombia last year, making one of my homelands amongst the most dangerous countries in the world in which to be an environmental activist; if it weren't for how all this is my home, I could also have just told her, given her the full list:

'------'

'Oh, I didn't mean anything by it!' she says to the silence.

If it weren't for all the other times that Jessicaaaah has been placed under the microscope and asked: what is this? How does it work? How does it use its hind legs — that accent? If it weren't for how being mixed in language as well as colour means that people read you according to what they're expecting to see — that you are made a chameleon at their disposal, and if I hadn't just realised when, and with whom, Jessicaaaah comes out; if it weren't for how obvious it now is that Jessicaaaah, in fact, sounds a lot like this lady — that she was modelled on her, even — I might have heard a poor choice of words in her question, instead of an interrogation. All she wanted was books for her friends.

'Well, we're all a bit of a mix, aren't we?', I say with Jessicaaaah's tongue swishing more Jessicaaaah-like than ever, in the same tone of voice that I say 'and the last three digits on the back of the card, please?' If not for all this, I might have asked her a better question, something to keep us both there, in the gap, before I made myself what I thought she wanted.

2.

This person comes into the shop not wearing a mask. Where in the crisis are we, me and her? Possibly it's July. Without a doubt, the shop where I work has survived thus far, and that's also a location to exist from, a vantage point which was never a given. Perhaps, many of us are less scared than we were three months ago, but angrier, and for others it's the other way around.

There's no one else in the room, and she clocks me as I step behind the counter, nods but doesn't say hello. I do, but behind the mask she might not be able to hear me. Bath is shaped like a bath, and although I like the idea of baths, I can't stand them for longer than ten minutes — too much of a lizard kind of blood, which soaks up the heat too quickly but also can't manage cold. One of my colleagues calls me 'the worst Viking in history', which makes me laugh, but also reminds me of watching *Lord of the Rings* and finding no brown girls in it, no way of dreaming myself into medieval versions of the North. What happens to our blood when it gets hotter?

She holds on to her bag, perhaps for purpose or because of feeling watched, the way I always do if I'm the only customer in a shop.

'You all right, can I help at all?' I say.

She snaps her head back toward me and then quickly looks away.

'Please can you take off that mask?' she says without looking at me.

'Sorry?'

'Please can you not wear that mask it's really triggering. I was gagged and it's triggering.'

She pulls up her hands, like blinds on each side of her face, in front of the graphic novel section.

'---------'

'It's the government guidelines,' I say instead.

She's crying now, but she doesn't leave. Instead, she flaps a picture book on the counter and begins to fumble in her bag. That automatic shop hand which belongs to me does what it knows how to do and scans the picture book, then types the amount into the card machine.

'People don't get it,' she says. 'It's like when they made gay people wear pink stars in Germany. I have a note here, from my GP.'

'What?' I say. 'Oh, no, there's no need.'

'I've been looking after this man on my street, shopping for people, I've been doing all of it.'

She wipes her eyes and looks at the counter. The transaction of money and goods takes place, through the little gap in the screen.

'That's really good of you,' I say, and try to look her in the eyes.

She's calmer now. She's survived this too. I do not think I was of much help in her survival.

'-------'

She leaves and I continue recommending stories that I think people might like, taking money for the books, thinking about how I did take my mask off when she asked me to, and that I'm not sure which way that plays, or who it helps in this economy.

3.

There's a story my mother told me at some point during my teens. It happened when we were kids and visiting relatives in Norrland, literally 'Northland', the north of Sweden. The few times we went up there, throughout my childhood, my sister and I loved it. In the summer, we swam, found stray baby birds on the meadows, and went fishing; in the winter, we rode sledges. Maybe this was the summer we found all those bird nests, and tried to put the birds back up in the trees. It could have been the summer my uncle let me have a go on his drum kit out in the shed. All these details are merely there to hold the memory in place, as if it needed anchors, as if it were unrelated to all that we are and have become.

At one point, my mother was playing with a small boy, a friend of the family. Who knows what kind of game they were busy with (was he on her knees or sitting on the floor), who was playing what, or if it was more like singing than playing (itsy-bitsy spider? How old was this child?). The whole game might have lasted ten minutes. In any case, the boy held out his hand toward my mother in the middle of their game. Maybe she thought he wanted to take her hand. She may have assumed he was making up new parts of the game as they went along, his imagination unfurling before her eyes and dancing with hers on this yard, so far away from the place that she, herself, grew up in. Before she became an economist, my mother had wanted to be a doctor. Because of institutional sexism (the university told her father as much), she wasn't accepted onto the course, so she studied speech therapy for a while and spent a few months working with children. Were they younger than this child? Did they have more or less of a language in common with her?

She held her hand out to him, but instead of taking it, he hovered above her skin. He touched a small index finger to the back of her hand, swept it carefully across that skin. Then, he held it up to his face, to see if something of her colour had come off.

'_________'

I don't know what my mother said.

'_________'

I can't remember what my response was when she told me.

Now that she's no longer here, the way I held (or didn't hold) that story is hooked like an extension cord into every experience she had of racism and xenophobia, that I don't know about. It's set alight every time I myself am un-belonged, made to feel free-floating without ramifications. It is what I have in place of every conversation we no longer can have about racism, and the different impacts it's had in our stories. When I was growing up in Sweden, the language with which to describe skin colour, to invite it into the whole of you without making people feel uncomfortable (because people were, in fact, uncomfortable), wasn't available. There were even fewer terms for our mixedness, or for simply being brown. At a pinch, you'd say mörkhyad, dark-skinned, but I suspect that, even then, this felt inadequate, like tiptoeing around a subject that needn't be tiptoed about, where those affected don't get to choose the steps. I had my body but no body politic in which to make sense of it. Neither, I think, did this boy.

I have to wonder if she laughed off the gesture or said something to the boy, if she (most certainly didn't) tell his parents. In my skin, by way of my hands, I make informed guesses as to how she might have responded.

4.

How does my truth react with someone else's to give us both severe headaches, make us both despicable people? The guy next to me on the DLR knows all about prams, but I'm the one pushing one.

'What's that about?' he says.

I was expecting this, but I'm also tired and flustered. Along with Adam and three friends, I've been pushing my pram all day with an oil barrel inside it, the words WHERE MY BABY COULD BE sprayed on the side.

'We've been to a protest,' I say.

When I should have said 'a performance'. That could have been the end of it.

'What sort of protest?' the man says.

'To support young people on strike. They're trying to get the government to act on climate collapse.'

'No,' he says, and shrugs like he's remembered something awful that happened a while ago, 'what's *that* about?' and he points at my barrel, which, for the day, I've named Meryl. He looks angry. The anger was there from the start, but I hadn't noticed because I was focusing on manoeuvring the pram with the barrel.

'It means,' I say, 'that the government is choosing oil and money over our children.'

He looks at the pram like it's someone's insides after a night out.

'That's wrong,' he says, and shakes his head. 'That's not right. That's *ugly*.'

Which, I think, is an interesting word to use. It points to aesthetics and taste. Art can be ugly and powerful. A good story, any story that says something worth remembering, always has ugly bits in it. Truth is so often ugly, and when it comes to climate collapse, which is supposed to be what this is all about, everything is hideous. It should make you want to throw up. But that's not what he is referring to.

'Are you familiar with the IPCC report?' I say.

'I don't need that shit.'

'Some of us are too afraid of what's happening to have kids of our own.'

'I know what's going on in the world,' he says. 'I have four children.'

At this point, our encounter has become a spectacle to the commuters. Considering my get-up (a black dress, tears of black paint like drops of oil down my cheeks), they might assume that the entire conversation is scripted.

'You're pathetic,' he says.

But it could also not be — it could also be a man (white) telling a woman (brown) that she's pathetic, on public transport, with the woman looking very uncomfortable, and no one stepping in, no one asking if she's okay. It could be the woman making the man supremely uncomfortable by pushing an oil barrel in a pram, dressed as if she's in a circus. One context doesn't negate the other, but the encounter is happening and this is how we respond.

'_________'

By now, Adam has managed to squeeze past a knot of passengers to come closer.

'Yes, pathetic,' he says, leaning closer to the man, 'as in worthy of pity.'

This doesn't work as he intended it to.

'You're pathetic,' says the guy again, 'that's why you don't have any children.'

It's a skill at the best of times, to talk to the general public during a protest, but this is no longer a protest — we are on our way home. He didn't choose this, he didn't ask for me to come into his commute, my anger, my grief, piled up in an intimate family accessory. If it weren't for the pram, we might have been better off.

'________'

But am I not also a member of the public? Isn't this crisis a public one?

'________'

If it weren't for the pram, he might never have spoken to me. If it weren't because I am myself terrified of not being equipped to be a parent, because his comment has triggered my own historic conflicts about sex, body, and birth, I would have told him to fuck off. If it weren't for how personal it all was, I wouldn't have taken it personally.

'You know you're being really mean?' I say, like a child in the playground.

And then we stand there, in silence, for the remainder of a train ride.

5.

Crisis, I read, is a latinised version of the Greek 'krisis', meaning a turning point in a disease, recovery or death, but it also comes from 'krinein', meaning to separate or judge. When does this happen in a conversation — in any gap between humans — what is made possible and when is that possibility forever lost? Until what point are we both able to respond, and to respond to each other instead of echoes of ourselves? In *Lady Chatterley's Lover*, D.H. Lawrence repeatedly refers to orgasms as a 'crisis'. It is hilarious to read this out loud, again and again. He never describes the recovery part.

One thing that seems to happen is this wanting to know, immediately — are you on my side? Please tell me that you're on my side in this,

'--------'

and if you're not, you're automatically on the other. The crisis is gone and we are back in the old order, the known, and failed, way to be. Between the two points lies the potential bridge. We have everything to lose in the encounter, but without encountering, we are trapped in sealed containers.

In April 2020, Arundhati Roy described the Covid-19 pandemic as 'a portal, a gateway between one world and the next'. She was writing at a moment when something unimaginable was happening, and anything seemed possible as a result. When does that moment end; is this still it? Are we still in the portal or have we come out on the other side, into the same tired picture, with the same deadly motions at work, to bring on new crises?

The idea of a crisis as a moment of opportunity is a fraught one. The word 'opportunity' itself so steeped in individualism, and narratives of winners and losers: grab it, seize it, make something of it (which are other ways of saying, make sure it stays yours and not another's). This isn't the kind of opportunity Roy writes about. Her description of crisis is steeped in pain and suffering; it is avoidable tragedy that has become inevitable change, in which the uncertainty may just allow for power structures to shift, perhaps to crumble. It's not an opportunity to be grabbed, but to be lived in, and shared, collectively survived.

Which pulls me back to the smallest crises, the minute, everyday separations when language and reciprocity breaks down, and where it continues to define our coexistence. These are moments when I change, or when small alterations might have happened, and didn't. The possibility for change resides, always, between the self and other, across the chasm from another. We have never arrived there on equal terms. We are there with immense risk, frazzled and bare, with different levels of vulnerability, unequally armed or not armed at all; we have come flayed and bleeding. Often, I want out as soon as I possibly can. Often, I want to stay, to remain in the risk zone without burning up.

Signal — a transatlantic response

One of my addictions is to childhoods. Those that come out sounding like: 'I spent mine in Fife' or 'mine smelled like gasoline and orange peel.' I collect them like my sister gathered wine corks in a suitcase. When they asked her what for, she said:

bara ha dem.

'You could make a floating cushion,' they said.
'You could make a curtain across your bedroom door.'
'You could get into the Guinness book of records.'

Bara ha dem, was all she ever answered.

My sister and I said 'I grew up in' and no one's attention stretched that far. There's not enough nerve to reach such disparate docks. My childhood was spent in splits, developed a posy of peeled endings. But to what extent is my childhood a part of this ending?

I'm addicted to the exercise of crafting a place out of this —

conjure a stillness out of my loved ones.

Because my childhood was spent with them, all I want is
just to have them.

OUT OF THE TUNNEL

Me

Pilots are trained in how to avoid 'cognitive tunnelling', otherwise known as 'inattentional blindness'. The rest of us are not. Does this mean we're worse equipped, less innately inclined, to spot life-saving connections? Inside a negligibly tiny space, surrounded by airtight metal and the intricately mapped highways in the sky, the pilot's awareness is supposed to reach outward, to the not immediately obvious but potentially key for survival. 'If a light starts flashing in the cockpit,' a pilot called Chris tells me, 'the natural tendency is to focus on that one threat until it's been dealt with. When lamps two and three begin to flash, you might not see them, because you're concentrating on the first danger you spotted.' To counteract that instinct, they are taught to sit back and, at least for a second, take a breath; to oversee the situation as a whole, access what they've learned, and then, with a view as broad and informed as is humanly possible, respond.

In an interview in 1999, poet Adrienne Rich spoke of the 'humanly possible' as an expression, questioning it in the context of the 'horrible culture of production for profit'. 'What's humanly possible,' she said, 'might be what we bring to the refusal to let our humanity be stolen from us.' Which part of these responses is human — which ones define us as the humans we are?

The pilot training, a seemingly specialist set of skills, is interesting to me for many reasons, only one of them being that I'm culturally addicted to flying but it now scares the shit out of me. Emergency signals are also flashing everywhere and all around: on the news, in my chat groups and social-media feeds, on the supermarket shelves, themselves shocked at being empty; the signals flash in different languages, with varying degrees of emojis and sass, depending on the subjective distance to peril, but undoubtably they're all relaying versions of the word 'danger'. They're all equally sure that this, this new thing, is the most urgent threat and we must deal with it first.

During the second UK lockdown in 2020, Adam and I attend a webinar about climate justice and the airline industry, during which Todd, a pilot, and Finlay, an aircraft engineer, speak. 'I want to talk to those guys,' I say to Adam, who's sitting next to me on the couch, a miniature version of our lounge among hundreds of miniature versions of hundreds of other rooms on the screen. We were supposed to introduce ourselves in the Zoom chat, and where we were, geographically, but I forgot to write 'England' next to 'Bath' and this provokes amusement. Why on earth would I assume that anyone knows where this Bath is — and that I'm not making a poor joke about life in lockdown — when I didn't know, less than a decade ago?

I have this desire to speak to them because of more than one vested interest. Partly, there's a fascination with the professions that quite literally brought me here, to Bath, a small city in the south-west of England in which some people walk around in Regency-era clothing and others sleep rough inside the nooks of every other shopfront on the high street, questions about how those working in aviation have facilitated my choices and I theirs, in the loop of luxury capitalism. By flying the planes, they also brought us to crisis and collapse, as little and as much as I did, in being flown. Then there's the other choice they made: a decision to leave aviation, and their selves as they were made by it — what it might tell me about what constitutes that self.

The Engineer

People who have been in the climate movement far longer than I have tell me that it was never this colossal; never did it have as much momentum as in 2019. When Covid-19 began to spread, our very essence (a movement: a collective of bodies in motion) became an impossibility. The physical incarnation of our collective power was criminalised for the sake of our own safety, and we could no longer hear each other breathe; we could no longer perceive that extra-sensory electricity (extra because beyond our individual peripheries, not because it couldn't be felt) of a gathering of people tearing at the status quo. What scares me is the time that passes — that while the government obsesses about getting 'us back to normal' ('us' meaning the economic machine that kills most of us), people continue to die from heat-strokes, from droughts and ensuing food scarcity. Ecosystems continue to disintegrate, making further pandemics more likely. Our physical togetherness has been outlawed to save us, but within the abysmal inequality of late capitalism, who really counts as 'us' — which self does this emergency response serve?

The day I speak to Finlay, there is a book about the interlacing crises on our kitchen table. In *Corona, Climate, Chronic Emergency*, activist and researcher Andreas Malm investigates the difference between government responses to the climate crisis and to the pandemic. 'Here is a factual property of global heating,' he writes, 'it does not appear out of the blue and then retreat to wherever it came from, as Covid-19 was expected to.' The sudden onslaught of Covid has made it fit more neatly into the category of 'emergency', but there is a danger that not only enlaces this one — it shares its roots, and exacerbates the suffering it causes. In La Guajira, Colombia, a region in the north of one of my homelands that I've never been to, the Wayuu people recently called on the UN to intervene in their fight against Latin America's largest open-pit coal mine. The pollution and influx of workers to the mine inflames the impact of Covid on the Indigenous population, whilst the virus in turn ramps up an existing environmental emergency, leaving people without water, for drink, for hygiene, for basic subsistence. There is no time to wait for this to be over, whatever 'over' might mean (it doesn't look like the virus will quickly 'retreat to wherever it came from'), and what if it's the perverse focus on one aspect of a larger crisis which obstructs a wider view?

Finlay is in Scotland. He's living at home, currently unemployed and dedicated to activism. His Scottish accent triggers my linguistic sponge, although it's been five years since I lived in Scotland. Because I didn't learn to speak in English, there's no natural, or unnatural, way for me to speak it — my voice does as Romans do, anywhere and everywhere, in a mix some people find curious, others take as a personal challenge to unpick. Now, I'm worrying that he'll think I'm doing it on purpose, so I watch out for certain words and, as a result, forget how to say 'how'. When Finlay trained as an engineer, he went in with the ambition of working in renewable energy, but upon graduation struggled to find a job in that sector. Instead, he was offered an internship then a graduate role as an aircraft-engine designer with one of the largest engineering companies in the UK, and the world's second-biggest aircraft-engine manufacturer. 'I just couldn't turn it down,' he says. He accepted the job with the hope of helping develop more sustainable engines. 'That's how I justified my decision,' he says, 'I was going to be changing the industry from within.'

Going through my notes later, I notice that the word 'justified' is underlined. I think I was wondering if Finlay used the same word at the time; if, when he was making the decision to take the job with a big polluter, he saw it as one that needed justification. It could also be my bias around the word itself, that I'm projecting, telling myself that writing about change, in the smallest way, somehow makes up for the big changes which seem impossible to push for right now, from our cluttered kitchen table.

I'd assumed that using less oil in an aircraft decreases the carbon emissions, but this is only true if the same or fewer planes are flown. A smaller cut doesn't spill less of someone's blood if it's one cut among a thousand. 'It's really about the price of oil,' Finlay explained to me. The people at the company he worked for 'are mostly good people trying to do the right thing, just not quite understanding what the reality of the situation is'. With more-efficient planes, flying has become more affordable; those who can, fly further, and more frequently. Although people had seemed interested in the sustainability group he set up at his workplace, it became clear to him that the employees were 'being deceived by the company'. 'They're told, you'll have a job, it's stable, but that's an illusion because either the industry, as it currently stands, is dead or the planet is.'

This false sense of safety is, evidently, not exclusive to aviation. The ice creaks under so much weight. It groans under our food supply chains, screeches around the basic integrity of our houses and coastal cities. 'It's fine,' governments say, 'we have ambitions to do something about the persistent sound in thirty years' time.' What's striking is that the history of aviation was written in the language of risk assessment; we wouldn't be here if it hadn't. They're still saying it will all be well.

The company Finlay worked for describe themselves as an 'environmental, social, and ethically sustainable business'. Their website boasts about how, altogether, their engines have amassed '100 million flying hours', a number I find hard to swallow, like leftover food which your body is telling you might have gone bad. For someone looking at this figure as a passenger, wanting to be reassured that they're in capable hands, the figure looks steady, it holds you fast with the brute force of quantity. If you're horrified of climate violence, it looks like the company is bragging about murder. On the other hand, I didn't even know that the company builds aircraft engines — I had, in fact, no idea who did, and that's also a question, isn't it? Who was carrying me, all this bloody time? Which one of these readers am I, looking at the 100 million flying hours?

I knew, well before talking to Finlay, that technology doesn't offer us a lifeline, not if by 'us' we really do mean all of us, which people by and large tend not to mean. I knew that it doesn't address the root causes of the crisis, but may well extend them further (where do raw materials for this green technology come from, if not from the Global South, now facing a 'greener' colonialism?), penetrating deeper, rotting whatever is left.

The aviation industry's obsession with tech fixes did remind me of other instances of dangerously selective attention. At a glance, the purpose of climate activism looks something like this: we need to halt further global warming by bringing down the emissions that contribute to it. Climate activists do this, largely, by demanding action from governments. This, in turn, can't be done without getting enough people (people with a voice) to see the climate crisis as a true emergency. This, essentially, means making people scared. It involves shocking those who need shocking out of torpor. In some ways, it's simpler: the alarm, the house on fire. They are images we understand, although so many struggle to apply them to our own lives. I started feeling uneasy about this narrative in October 2019, during Extinction Rebellion's second large protest in London. One morning, a group of activists climbed on top of a train at Canning Town station. This led to chaos on the platform, with stressed commuters, desperate to get to work, shouting abuse at the protesters. One activist was dragged onto the platform and attacked, while other commuters stepped in to protect the protestors. I was in Trafalgar Square when it happened, handing out water biscuits to people who were locked together blocking roads, and I remember wondering, why *there*. A majority of XR UK (which was, and to my knowledge remains, a decentralised movement) disagreed with the action, which had mainly affected working-class, mostly non-white, people, but this was a campaign that grew, to a great extent, out of grief, fear, and anger. Its focus was overwhelmingly on 'waking people up'. Those protesters had done what they thought was needed to puncture everyday life. That urge, it seemed, had excluded any other consideration.

It's been almost a year since then. We've all moved online — the meetings, the workshops, the family dinners. Sometimes, on a weekend, people go out with placards saying #NoGoingBack. It's an acknowledgement that normality outside of corona was already unacceptable, but who's holding the placards? Who's able to go out protesting at this time? Again, how far does the crisis stretch — how small is this tiny space?

The Former Pilot

Todd wanted to fly aeroplanes since he was five. 'Flying fighter jets for queen and country was the dream,' he tells me, and he does not smile when he says it. He's a few years younger than I am, in his early thirties, and there's sun bleaching his screen when we speak. He describes his childhood as a bit rough, growing up on a working-class estate, and always with flying as the next step, as well as the ultimate goal. In order to get him through the training to become a commercial pilot, his parents remortgaged their house, in the wake of the 2008 financial crisis. Two and half years into a job with a commercial airline, he was 'medically grounded'. Later, he would be diagnosed with Lyme disease, which he contracted from a tick in Richmond Park. 'They're becoming more common with the milder winters,' he says. During this time, he began to attend occasional meetings, to get involved in protests, whilst remaining part of his professional community. Until the pandemic, he says that he could still imagine working, perhaps not as a pilot, but within aviation. Now, during this hypervigilant time, which looks like sleep for some and is life-threatening for so many others, he describes his relationship to aviation as a 'love affair in which [he] was betrayed'. Having taken a step back, having paused for that breath, he began to question everything he'd ever wanted.

I'm thinking of the almighty words: the Writing, as it took a plunge into my life and never popped back up. One day when I was seven or eight, I happened to write a poem which my teacher did a song and dance about, and I suspect that on this day I said to myself: this, this is what will make it okay to be you, in spite of every other way in which being you needs urgent improvement. There have been times when I've looked up and spotted alternative lives, things that need doing. I could, potentially, help with these things, but they're not part of the story I decided on when I was eight. This too is a sort of tunnel vision: a narrow view of yourself in a reality that needs you to be so many things, to allow and invite every facet of you that may bring you closer to others.

In between interviews with aviation people, I'm now involved in organising yet another online meeting. This one is about how to make social justice an explicit part of the principles in our local climate-activist group. It is the autumn after George Floyd's murder and the ensuing Black Lives Matter protest wave. The city we live in is overwhelmingly white, a place where many are worried about 'environmental issues', but which is also used as a weekend getaway for Londoners with disposable income. The city voted against Brexit, but I often do wonder why, for its own freedoms or for everyone's? Would they be content if those freedoms stopped at their threshold? At first, it appears that those at the meeting generally agree with the notion of social justice being intrinsically linked to climate action. There is effusive nodding when we talk about communities at the front line of climate collapse. Still, we fail to make any of it official — we fall short of making any concrete commitments, of saying, loudly, that 'racism and climate collapse are one and the same'. Someone argues that we need to remember we're an environmental organisation. The climate crisis needs to remain our priority, someone else argues, and it's clear that they're referring to the atmospheric science of CO_2 and methane, not the climate crisis as a whole — not the bite it has on people's lives. Anti-racism need not be an explicit ambition, as long as we're inclusive in our work, they say. Emphasising the roots of climate collapse somehow derails the urgency: we don't have time to pay attention to how this happened.

In the moment these things are said, the people for whom the system never worked are discounted, made invisible once more. This is ancient, timeworn history, of course, and the racism within the environmental movement goes the other way as soon as guilt sets in. This summer, at the height of the Black Lives Matter protests, essayist Mary Annaïse Heglar wrote about how strange she felt, as a Black climate activist, hearing white people argue that climate action had to be halted in favour of supporting Black people, as if people of colour weren't disproportionately affected by climate and environmental collapse. 'So it's not just time to talk about climate — it's time to talk about it as the Black issue it is,' Heglar wrote. Climate collapse has been a racist issue for at least five hundred years. Colonialism made it so.

And none of this is new, and always new for someone. In this group, I'm one of two people of colour. As an ethnically mixed, queer person in a hetero relationship, I often experience myself being what people want to see — my body and outward presentation reflect back what makes them most comfortable. To some extent, we are all adapted in each other's eyes. People prioritise one identity or another depending on which one makes most sense to us, which self in the other could provide safety and make communication possible. With power structures being what they are, whiteness and heteronormativity are so much more often than not the framing narrative, the way things ultimately land. I've always been made to feel welcome in this group, always listened to, but it's equally true that an inability to see a bigger picture itself excludes, as does the erasure of someone's multifaceted self. Telling any person of colour that anti-racism needn't be explicit in climate activism erases them from that activism, as does telling queer people that LGBTQ+ rights needn't be an official part of climate justice, or disability rights for someone with chronic illness. The bigger picture, in fact, involves both who we are and how we are tied into this crisis. Crisis response and identities emerge as intimately linked.

Which dangers are being missed, if the focus lies on only one level of threat — only the one flashing light, noticed through the bias of particular identity strands, certain affiliations? What kind of tunnel vision is this, which allows people to believe they're fighting the climate crisis whilst only fighting one symptom of it?

The Pilot

But how to *see* those '100 million flying hours'? The day of my conversation with Chris, I visit flightradar24.com, a website which tracks the number of aircraft in the sky over any particular part of the earth, at any given time. I haven't looked at it since the start of the pandemic, and the number of tiny yellow fireflies on the map remains astounding. Airlines are screaming for subsidies, the livelihoods of thousands are at risk, but the planes still appear legion. They twitch slightly, chomping through miles, seemingly nose against nose, but really so distanced, in a meticulously planned net. I remember looking at the map of routes in different airlines' flight magazines as a child, this intricate web, and not once feeling a sense of dread.

'How's it going with your pilots?' Adam asks. 'It's two pilots and one engineer,' I say, 'and none of them are doing those things right now.' This makes me question how I think of them. Once you know someone's a pilot, you think of them as a pilot, perhaps more than with most professions. A doctor is a doctor in a way that a window cleaner isn't only a window cleaner. Late capitalism does this to us: it encourages us to identify with our jobs, with privilege and status as the main driver. An article in *The Atlantic* calls it 'workism': society holds up identification with one's job as something to strive for, as opposed to work for the sake of survival. No wonder that it's difficult to let go of when you realise that there's something terribly wrong with your job. Your identity has been steered toward this narrow path; letting go becomes life-threatening.

On a Monday afternoon in early January, I ask Chris if he can remember a time when flying wasn't the focus — the non-negotiable core of whatever else was going on in his life. The laptop floats on a sea of tissues. If someone walked in, they might think this is a therapy session and I'm bawling my eyes out, but it's only a cold I came down with on Christmas Eve — I've taken two tests but still am not sure it's not Covid. The official list of symptoms is one thing here, and one thing in other countries, as my dad reminds me regularly. If it's not on the list, it isn't a symptom of Covid, but who decides what goes on the list? How are the red flags agreed upon? Chris takes a while to answer. It even looks like the question bores him. 'At some point,' he says, 'I wanted to be a vet.' I tell him that at some point, I wanted to train dolphins.

Unlike Finlay and Todd, Chris is still employed by an airline, which is why we're not using his real name in this piece. Climate activists aren't looked on kindly in the industry. So far, he's managed to keep his involvement under the radar. When we speak, he's just been called back from furlough, but not yet been asked to fly again. A while ago, he tells me, he had to attend a lecture about threats to aviation. Climate activists and deportees were mentioned as two of these. 'In what way are people who are being deported threats?' I ask. 'It's what they might do during the journey,' he says.

Again, as so often these days, I get a feeling of standing at the bottom of a well, discerning only a small circle of sky. There's an immensity up there, but I can't see it, can't make it any larger. I press my hands against the wall and feel rushes, courses, and a pinch, but I am in this well and the depth prohibits any wider vision of how exactly that pain connects to mine.

The lecturer, Chris says, also mentioned James Brown, the Paralympic gold medallist who climbed on top of a plane at London City Airport in October 2019. A few of Chris's fellow pilots laughed at this. One of them said that if it happened to them, they'd probably fly off with the protester on top of the plane. 'This is what the naughty kid at the back of the class felt like for all those years,' Chris says, because if you no longer belong, you become dangerous. Paying attention to a greater danger, you run the risk of yourself becoming a risk. Todd told me how it had taken him eight or nine months to leave the social-media forums he frequented as a pilot. When he posted about the climate crisis, most people were casually dismissive; some were outright hostile. In pilot threads, climate activists such as Greta Thunberg were mocked and presented as pin-ups. Eventually, he left for the sake of his mental health.

What is humanly possible?

I don't feel like the naughty kid at the back of the class. The people with whom I disagreed about how to take action never mocked me, and because I am mixed, this grants me the privilege to pass. We are all trying our best to respond. Still, I am caught in-between, because my fears didn't fit. My nerves flailed and stretched in too many directions. They strayed out from the track we were so urgently on. It's terrifying that even within groups that so desperately care to preserve life, this happens so very easily.

Multiple-Issue Lives

The last time I flew, I tried to remember what it used to feel like, and couldn't. I pressed a hand against the armrest and scratched its pockmarked plastic, knowing it, and rejecting it under new premises. The airline showed an advert for its environmentally friendly food and ambitions to cut down on plastic. As the engines started, the sound entered my body as something perfectly within the order of things, but my brain received it otherwise. It felt as though my nerves were no longer inside that aeroplane but further out, with the harm the plane causes. I don't want to say that they had lengthened, or grown, but that they were never confined to my body, and I had previously only perceived a fraction of them — that it's the image of myself that has changed.

Phantom limbs cause real pain. One explanation for this, proposed by pain researcher Ronald Melzack, is that pain is an experience, a subjective response that happens in the brain, rather than a reaction in the periphery. Because the brain has an idea of where the body ends, it continues to feel a missing limb. In other words, pain is linked to how we see ourselves, and where we see ourselves ending. If fear is, among other things, the premonition of pain, what does pain's subservience to self-image tell us about fear? According to one image of myself, there was nothing threatening in that cabin. Looking strictly at the faces of passengers, at the attentive gazes of stewards and my body strapped routinely to a seat, at everything in its place, threat could only come from outside of the system, in the form of an anomaly, a mistake. If I saw myself as more than this, as part of the whole that is being destroyed, the cabin itself was the threat — the harbinger of pain. The plane took off. My nerves kindled, aching and burning but not snapping. I was now, it appeared, afraid of flying.

If we can change perceptions of where we end, how much more will we be able to feel?

According to a study of airline accidents in 2005, stress impairs how a pilot appreciates information in the cockpit, their 'cue perception'. With greater stress, there's less attention to go around, and certain cues are left unattended in service of greater efficiency — of getting things sorted quickly. 'However,' the study says, 'some of the cues deemed irrelevant are sometimes relevant, and the "efficiency" achieved comes at the price of embracing an incorrect interpretation of the unfolding scenario.'

Do we simply do this, when we are afraid? And many of us are so very afraid. Does our field of vision, and feeling, immediately shrink until we have only attention for the danger with our name on it? Until we can only interpret the scenario through a narrow lens?

In a moment of crisis, such a theory of human behaviour lies close to hand. It becomes easy to think of it as simply what happens to us, to revert to finite definitions of 'human nature'. What is then ignored is the vested interests present in this version of what humans are. Someone stands to gain from you identifying with your job only, and not with the rest of your humanity, which may challenge the very premise of that job. Courting single-identity lives, striving for inclusion in single tribes, benefits capitalism because a limited self is easy to control in the service of unlimited growth and accumulation of power. As the pandemic is making abundantly clear, it's an immense privilege to be able to take that crucial breath, to have the time and a moment's peace, the access to enough stories, to take thorough stock. When it comes to big life choices, few people get that chance; they are too busy surviving. This plays into the hands of the few, the powerful, and those who'd want us afraid. It's easier to crush a single-issue movement than a movement of endlessly entangled threads, as multifaceted as we are ourselves.

Returning to the 'humanly possible', here's Adrienne Rich again: 'Human beings aren't merely determined by capitalist production — Marx never said that. These are conditions "not of our choosing" in which we can make history.' Tunnel vision may be a human tendency — that impulse to get stuck on one solution, which also implies one vision of ourselves and others — but human beings have many instincts, and we're not defined by a single one.

It makes sense that those industries with the most to lose if we do what needs to be done to avert the worst of global heating — fossil-fuel companies, aviation — would become breeding grounds for judgement and paranoia. This is part of the same narrowing, an amputation of feeling and of being. Those industries — any group, any tribe — are still populated by people who are always more than one thing. Finlay, Todd, and Chris are all involved with an organisation, Safe Landing, that works for a just transition for the aviation industry, championing the rights of workers in the change that will inevitably come. In that sense, their selves are still sewn into the fabric of flying. When I spoke to Chris, he hadn't decided what to do and didn't know how he would feel once he sat inside a cockpit again. I asked Todd if he remembered what it felt like last time he flew a plane. He said: 'it's me at my best.'

Writer Amin Maalouf has described a person's identity as 'a pattern drawn on a tightly stretched parchment. Touch just one part of it, just one allegiance, and the whole person will react, the whole drum will sound.' Our bodies and ourselves are the parchment, woven through with coils of feeling. In a complex system, no corner is separate; there are in fact no corners, no real endings at all. Fear may increase the risk of narrowing our field of vision, but that fear, for the world seen and unseen, that urge to preserve and to respond in its defence, are just as much a part of the drum.

Whenever I feel pushed toward a sharp edge, to define what I am 'about' once and for all, to be anything less than the mix and in-between-desires that I am, the words that come to mind are always those of Audre Lorde. 'There is no such thing as a single-issue struggle because we do not live single-issue lives,' she wrote in the essay 'Learning from the 60s'. This text was delivered as a talk in 1982, six years before James Hansen gave his speech about dangerous climate change to the US government (which in turn marked the beginning of the fossil-fuel industry's full-frontal attack on climate science), but the words appear as a blueprint for the kind of response the climate crisis demands. Elsewhere in the same text, Lorde writes: 'The answer to cold is heat, the answer to hunger is food. But there is no simple monolithic solution to racism, to sexism, to homophobia.' Neither is there for the climate crisis, because all of the above, along with an economic system which protects corporate greed, helped create it. It's not easy to recognise this, when we're afraid and seemingly quicker fixes are offered. Neither is it impossible. Because we are always multiple, responding in multiple ways is in fact more humanly possible — it's the kind of action that makes us human, and cares for the human in others.

One way of classifying nerve endings is by the way they listen. It is to say: you are an ear to the wall or one to the belly. Another kind of nationhood. There are the ones

that cry me   the ones that cry not-me,

meaning the ones that cry world?

Another is to name them after what they feel: a fall and the cut-throat layers of snow, a love's name for you and inherited words for a terrible cold, los ojos en la paila. I love you for what you love, not where you began. Pain and fear permeate the in with the out, the else into us.

What kind of nerve endings detect the loss of origins? Tell you to wake to this kind of fever?

Another friend asked: 'what grief isn't personal, anyway? It all happens in the same brain.'

FREAK AGUACERO

I

This is how I learned about the climate in Colombia: starting out with a plain template (didn't our country look like a bloated starfish, perhaps one in stilettos?), we were instructed to populate a series of maps. These would later be bound and graded by our primary-school teacher. Each map began with the same contour, until the line could be traced almost without looking; each would serve as a face the motherland wore, a distinct way in which it was la nuestra — ours — as the advert for the Colombiana soft drink insisted that it was. Ours in what way, exactly? This was in 1996, in Bogotá, the city where my mother was born.

In January 2019, just under ten months after my mother died, my sister and I go to Bogotá together. We go together, for the first time as adults, but we've been adults for a good while already. Is this why everything feels so late in the day, so much past its use-by date? The nativity scene is still in our grandmother's fireplace, replete with miniature holy men and sheep, a leftover from the latest major event we missed.

The whole of the first week is spent in our grandmother's flat, occasionally taking her out for sun in the courtyard and, once, taking ourselves to a bakery across the street. When we want to go out for groceries, our aunt, our older uncle's wife, asks if we can please wait until one of them takes us. We have two uncles on our mother's side, and the younger has insisted on staying with a cousin, so that we can have his room. He's left us an impossibly thick duvet, handmade by a former girlfriend. Either we are being looked after because we don't know how to do it ourselves here, or here, this is how people look after each other. Although we used to come here at least once a year throughout our childhoods, and although my sister was born here, we only actually lived here for three years in the nineties. Our mother wanted us to know this as a home, to have this place sewn into us. The celador who guards the block where my grandmother lives says 'hasta luego, niñas' ('niñas' sounding so much younger than 'girls') every time we leave the building. We are both in our thirties.

This is how the climate in Colombia was taught to me: by proxy. Planning work, and estimating how long a certain piece of work will take, has never come easily to me, and the night before the map assignment was due, my mother and I sat up late. We watered the country with rivers, from thick blue ribbons to the hair-width of creeks: 'El azul de la bandera es por los rios, niños' ('the blue of the flag is for the rivers, kids'). We planted vast areas of rainforest. I say 'we', but mostly I stood next to my mother, who sat hunched over the desk, a side view from on high of her working right hand, drawing what was supposed to be tiny cattle. At the end of it, we had a folder — you could flick through it quickly and see the nation unfurl into separate layers of skin, a multiplicity of threads and wires stitching the land. There's no way anyone would have believed this was done by a child. The maps, in any case, would later be doodled over by a three-year-old, the son of a family acquaintance, who snuck into the room my sister and I shared, in search of who knows what.

On the first Sunday of our visit, my mother's younger brother, who travelled around the country with my parents in the early eighties and introduced me to Celtic music when I was fourteen, takes us for a walk through La Candelaria, the colonial centre of town. It's a touristy place, but also a heart of a kind; not knowing it leaves us heartless. The day begins with the usual smog lid on, the school buses exhaling like exasperated whales at every corner. My ears recognise this sound as the prelude to carsickness — mine and that of other uniformed schoolchildren in one of the world's most congested cities. This is only good; in the business of recognition, nausea is an expert.

My uncle parks the car close to the Hotel Tequendama and we walk uphill to see the view from near the planetarium. When in a certain mood, the sun in Bogotá chomps down on your scalp and takes bites out of the thin skin under your eyes. 'El sol aquí tambien quema,' our mother used to say: 'the sun burns here too'. No one thinks of sunbathing in Bogotá. The sun wants you to know that you're edging in on its territory. Other than that, nothing about the explosion that is this city, all ochre and uneven roof tops, makes it tangible that it sits 2,640 metres above sea level. It's a colossal basin filled with almost ten million people and still growing. It's able to breathe but not feed itself. The capital's dependency on neighbouring regions makes food scarcity one of its main vulnerabilities in the face of climate collapse.

On the long flight here, I paid for on-board Wi-Fi for the first time ever, so that I could look up Spanish terms related to global heating. *Cambio climatico, industria ganadera, puntos de inflexion*; I didn't have words for the end of my world, in the language in which it began. With a handful of terms, I then try to explain to my uncle, repeatedly over the course of two weeks, why I've gone vegan. 'It makes sense, all those cows and the methane,' he says, jokingly. Cows, I think, and try to make the jump from that to 'urgent system change', and the between-moments desperation that makes me want to block roads back home, even though I'm not at this point a UK citizen and could have that home revoked. My teeth desert me for the rest of the conversation. What do I want him to say? That he will be going vegan too, at the age of fifty-nine, in a country where vegan food costs a fortune or is non-existent? Or that he'll be joining a protest to demand action on climate change? What is this 'system' that I want to change, anyway — the one which allowed me to grow up on two continents, or the one in which five people from three generations share a small flat? Both are part of the same system and it gave me my life. I'm the result of my father, who worked for a Swedish company, moving to Colombia for a few years and meeting my mother there. Regardless of how much my mother's family is mine, that history, the imbalance of North and South, of coloniser and colonised, puts me on the other side of a global divide.

After spending an hour at the Botero Museum, we walk down Calle 11, toward Plaza de Bolívar. The sun has given way to a chalky pressure over the eyes. This is only one of hundreds of squares in Colombia named after Simón Bolívar, El Libertador, but the only one where the assassin of Jorge Eliécer Gaitán — a progressive politician and leader of the Liberal Party — was dragged by an angry mob in 1948, sparking the period known as La Violencia. As if by giving the name to a chunk of time, violence might have been confined to it. While he was still alive, our grandfather dropped hints that we may be related to Gaitán, but when I mention this now to our uncles, they laugh. There are too many Gaitáns in Colombia for this to be worth pursuing, and too few records for it to be possible. All of which sounds like an excuse not to dig into things. Received belonging can be alienating, as well as comfortable.

One of the maps was dedicated to mountains: three ranges born out of one, as if a giant hand at the border with Ecuador filters the Andes between its fingers: la cordillera occidental, la cordillera central, la cordillera oriental. At the very top, there's la Sierra Nevada de Santa Marta, which I thought looked like a spot on someone's forehead, snow at its peak. The thirty thousand Indigenous people living there, the guerrilla and paramilitary groups fighting over the mountain for decades, weren't to be included on the map.

Our version of the ranges ended up looking more like slug trails, or the veins on the back of my grandfather's hand. Luckily, I didn't tell this to my mother, who must have been exhausted by this point. The maps had been my responsibility, not hers.

The main attraction on Plaza de Bolívar seems to be a pair of llamas wearing pink ribbons. One of them is being encouraged to dance. At some point between a quick look inside the cathedral and us finishing our corn on the cob, it begins to rain. Precisely because it comes out of nowhere, there's nothing strange about this rain. Rain in Bogotá often behaves like this, pouncing on your day like a raiding party. My sister, my uncle, and I run to wait out the aguacero underneath the awning of a discount shop on Carrera 7, alongside at least twenty others, but soon decide that it's too crowded and start walking toward the car. That word, aguacero, always puzzled me: it sounded like agua-cero, zero water, but refers to the opposite: a ferocious downpour. Aguaceros never remain aguaceros for long, which is what makes them aguaceros, as opposed to any other kind of rain, to be expected anywhere else.

Where Carrera 7 meets Calle 19, my uncle cuts a corner a few feet in front of me and is slapped over the chest by an open-handed wind. It throws me backward and makes him hold on tight to his jacket. My sister's face stretches out to its edges, giving her a stranger's grin. She was born with an ever-so-light line down her nose, a birthmark which our mother said was from the factory where they put dolls together, and now this line risks being cracked open. In Edinburgh, or Göteborg, a wind like this would be native. It would know its way around the city and understand its routines. Here, its shriek is incompatible with any other sound. We've been plucked up and deposited somewhere else, where winds such as these exist, and where downpours include such winds. The wind, being not of here, has made the weather itself alien — it's even changed the aguacero, derailing it to the sides.

This is it. I look around to see if anyone else gets it, that this wind is a sign of certain collapse, of higher temperatures and changed currents. You freak, I think, I know what you are; at least I know this. I am, weirdly, comforted.

Next, there was a map of major cities. Colombia's capital is located smack in the middle of the Cordillera Oriental, in a valley called la Sabana de Bogotá. It was legally founded in 1539, having fulfilled the Spanish crown's requirements for a city. The name derives from Bacatá, the capital of the zipas, the chiefs, of the Muisca Confederation, which was burned by the Spanish the year before. Its altitude places the city in the category of tierra fría, as opposed to the tierra caliente of the coast and of lower altitudes. You wear different clothes to tierra fría and to tierra caliente, but neither of them have anything to do with a deadly hot earth. Aged ten, the 'savannah' in Sabana de Bogotá only made me think of *The Lion King*, which was also something that happened in the mid-to-late-nineties, along with the expansion of guerrilla territory, increased regularity of paramilitary massacres, and Plan Colombia, the specific local strand of the US war on drugs.

II

Upon returning from our trip, my sister and I speak to our family in Colombia once a week. We take turns calling our grandmother, me to read her stories from a collection by Isabel Allende that we started when I was there (the raunchier — the more wooing there is — the better), and my sister to play traditional vallenatos (heavy on the accordion) over the phone. While my mother was alive, we spoke to our Colombian family almost exclusively through her. She isn't here now, to keep the connections live. It's on us to make sure they don't drop.

On the phone, when my aunt asks what my partner and I are up to, I tell her that we're busy, with 'work and climate things'. I never go into too much detail. It's not that I don't want to tell her, but my willingness to explain, and to be questioned, doesn't stretch far enough.

No map, as part of the exercise, was dedicated to the Indigenous inhabitants of the region we knew as Colombia, neither what they were before the sixteenth-century genocides, or what they are under the current era of colonialism.

There are openings, in the months that follow, chances to talk about the climate crisis with my family. My older uncle texts me to say that what my partner and I are doing — the protests, the meetings — is good. 'Either we change or we all go extinct,' he writes, and him using that word, *extinct*, makes me twitch. He's used a secret code, meaning that he knows; our fears exist beyond the particular climate of activism we've ensconced ourselves in. Somehow, I must still be thinking that if certain people, people I trust, don't take it seriously, it must not be that bad. My younger uncle texts me to say that I left behind a climate-protest flyer on his bedside table (I was using it as a bookmark); does he assume I left it there on purpose? I don't ask. He says it looks interesting and that he'll have a closer look.

The river map turned out quite well, I seem to remember — one blue stripe connected to another, nerves made terrestrial. 'Los ríos de Colombia son muy importantes para el comercio,' a teacher said, but I only remember walking into one actual river. It was during a visit to my mother's aunt and uncle's finca in Los Llanos, where it was so hot that our mother sat up through the night waving wet rags in our faces. There was a story about the son of a caretaker who almost got eaten by a snake on the banks of that river, and I imagined the size of such a snake when paddling in it. I didn't consider the relationship between my family and the caretaker, his son, and the land, who owned it and who never had because the notion of owning had been forced down their throats. I drew some of the river map myself, but I don't know the name of that river, or how it's doing these days.

I begin to research the impact of climate collapse on Colombia, and on Bogotá specifically. Perhaps, knowing about local floods and areas prone to desertification, I will have a leg to stand on when talking to my family about climate collapse. Really, they never ask me about my legs, except when I've hurt them. They've never questioned what I'm doing, yet I seem to feel the need to defend myself. I learn that Colombia, being a country where the vast majority of the population lives either in the Andean region (with risks of water shortage and landslides) or along the coasts (where the water is rising), is described by the UN as 'especially vulnerable to climate change'. The country's infrastructure and 'precarious settlements' are part of what make it 'especially vulnerable'; especially in relation to other countries, and especially for displaced and poor populations within the country itself. As for Bogotá, there's the risk of dengue-carrying mosquitos making themselves at home as temperatures rise.

Most of this is accessible climate science, the kind that even I — a 'book person' — have come to deliver in lecture halls as part of this 'doing something'. It isn't really what I'm looking for. I spend hours searching for mentions of storms in January 2019: a particular wind that didn't behave the way it was supposed to, and that must have been a sign, but none of the results are satisfying. None of them confirm my experience of an aguacero that no longer felt like one.

In April, three months after our visit to Bogotá, my partner is arrested in London for refusing to leave a protest site. On the train home after his release, my younger uncle happens to text to ask how we're doing. I tell him what's happened, because I'm still high on adrenaline, and, probably, admittedly, a little proud. 'Are you sure this hasn't gone too far,' my uncle replies. 'Lo cortés no quita lo valiente,' he writes, a saying my mother also used which literally means 'the polite doesn't take away the bold', and I reply that this isn't about being bold, but about what's necessary. Sometimes, impoliteness is more than necessary, and he should have seen the protesters. I wish they'd been less polite.

Although of course it has to do with boldness, why else would I feel pride? Although a brown person, I was surrounded by mostly middle-class, able-bodied white people during that protest, and this brings with it a certain, more restrained, police response — a safety which cannot ever be taken for granted in Bogotá. Where does this leave my politeness, or my pride? It's a different climate here, I add, to reassure him that as far as the protests go, I'm not in any immediate danger.

We also had to memorise the names of each region. Most of them, by the way, were only familiar from the televised Miss Colombia beauty pageant. I knew that my grandmother was born in Santander, and I recognised the department of Huila because my parents had pictures from San Agustín, a pre-Columbian archaeological park. Eso está lleno de guerrilleros, my mother said, which meant it wasn't a place we could go anymore, in the mid-nineties. What these guerrilla soldiers did there, or why they did it, wasn't part of the formal education about the climate in Colombia. My sister and I knew of the region of Guajira as the setting of the telenovela named after it, but were oblivious to the opencast coal mine which, along with droughts, would come to make life nearly impossible for the Indigenous Wayuu population.

I ask my aunt over text: what kind of changes has she noticed in the climate over the last, say, five or ten years, and she tells me about the birds. Some have disappeared from Bogotá, whilst others, which you never saw at this altitude before, have started frequenting the city. What about the weather? I ask. The wind, specifically, and what about the rain? I send the same request to both my uncles, and instead of a reply, my younger uncle calls me. We talk about his love of hot weather and my uneasiness with it. His dream is to build himself a house with a small pool close to one of the villages about three hours down the mountain. Isn't he worried about it getting too hot? His answer is neither a yes nor a no. He talks about designing a cool-enough house (he's an architect by trade), with a wall around it. 'Donde me dejen tranquilo,' he says. Someone can throw food at him over the wall, he says. This would be his life, and it would be fine, as long as no one bothered him.

Talking to my family about climate collapse, I realise that it's not that they don't agree with me on how bad it is, but that they

— like me — don't know how to respond. A response to future threats always depends on historical and present dangers, and living your whole life in a country with over fifty years of civil war makes mundane tranquillity — simply to be left in peace — a priority. It's not just geography that separates us, but time, a generational shift. Because of all of the above, I've had peace, which afforded me the space to worry about the future. This seems obvious, but when it's your own family, how do you best care?

I mention the aguacero with the strange wind. Oh no, my uncle says, you get that kind of wind every year in Bogotá around that time. It's because it's surrounded by mountains. 'Eso aquí es así,' he says, which is what he's always said about the insecurity, about not being able to trust people, about always calling taxis instead of grabbing one off the streets, and about government corruption. When he refers to people taking the piss as soon as you turn your back to them, he says 'eso aquí es así', as if he's given up on things ever being otherwise. What's more, I don't know enough about the way things are here to be able to tell him it's not too late. It sounds vaguely like when people say 'it is what it is', which Adam dislikes so much he's made me promise never to say it.

Possibly, this is what I wanted: for someone to tell me 'you were right.' Which would mean: 'You know this place. You know it well enough to see it change.'

If someone we didn't know asked us for our names, my mother said, we should always give our Colombian surname. The worry was that if they knew our father's Swedish surname, we'd be at risk of kidnapping. 'They' were the guerrilleros, but it could also be someone spiking your drink then leaving you in a ditch, which happened to a colleague of my father. One day at school, we were all ushered to a classroom and shown a documentary about the drug trade. It presented us with a clinically sparse room and a dead body in the middle, opened like a tamale, being emptied of packs of cocaine. I asked the teacher if I could leave, knowing I'd have nightmares, and the teacher said I couldn't, because this was very important. I could hold my hands over my eyes, if I wanted, but I had to stay in the room, she explained. That way, at least in theory, I would know what was going on.

III

When I've stood in front of a group of people talking about three, five degrees of warming, or stared at a camera during a protest, willing it to see enough fear, I know that, inevitably, I'm speaking from a distance. It's a gap similar to the one between me and the maps, me and the drug-trade documentary. Privilege filters climate fear; inequality makes it a warning rather than an experience, theory as opposed to lived reality. This doesn't void or lessen anyone's fear, but it does make a difference, and that difference carries through into our reactions to a crisis which is global, but always felt locally, and in endlessly different ways. Writer Sarah Jaquette Ray has asked if so-called 'climate anxiety' is really just 'code for white people wishing to hold on to their way of life or get "back to normal"'. Each of our 'normals' is characterised by vulnerabilities, by what may or may not happen to shatter the sky above us and ours. My 'normal' never had the same everyday risks and volatile days built into it. Although a global crisis, climate collapse can never be the same crisis globally. When we talk about our most existential fears across class, generational, or geographic differences, we're expressing very different wounds, and different abilities to tend to them.

It wasn't a lack of vocabulary keeping me from speaking to my family in Bogotá; it was guilt. When I think of the Colombian part of my childhood, I feel guilt because of its privileges — the private, bilingual school it happened inside, the Swedish passport that framed it, the shelter from violence — but there's also a guilt because of everything I missed when I didn't live there: a war, a peace process, what comes after. Finally, there is guilt because now I'm here and they are there, and because I don't know enough about their life to say what's best for all of us, even when this thing, this many-headed horror-show, will come for us all.

The map representing Colombia's climatic zones looked like someone had splashed five kinds of oil paint on the country's chubby body, then smeared them with a finger. Mine, my mother's finger. It was difficult to get to grips with. I stared and stared at it for a long time, trying to imagine what it felt like to live your life in each, to be me in each of those places.

A popular way of explaining the difference between weather and climate is through the allegory of personalities. If weather is a mood, then climate is a personality, it is said. This makes me think of the years when we lived in Colombia and my father suffered from chronic stress and road rage. Anyone who met him, for the first time, coming out of another six pm traffic jam, would have made certain assumptions about him as a person. We knew he wasn't like that, not normally, because we'd known him all our lives. When I ask my family about the weather, I say, 'que tale el tiempo?' which makes the word for weather the same as the word for time. Rather than confusing things further, this may work as a reminder, that in order to truly understand its climate, you need to give a place your time.

In late April 2021, protests break out all over Colombia. They started in 2019, but were interrupted by the pandemic. They are connected to the government's handling of the Covid-19 outbreak, but also to the broken promises of the 2016 peace accord (a watershed event I watched from afar, without understanding its metabolism), and to so much of what went before — at least half a decade of armed conflict. Between April and early June, sixty-eight deaths are documented by Human Rights Watch, many of them at the hands of the ESMAD, the police anti-riot squad. If I lived there, would I be part of the protests? If I was me, but there, would I develop entirely different ways to respond?

Instead of asking my family about what's happening, I've been trying to make contact with Colombians in the UK. It's somewhat counter-intuitive, not only because I always resisted connections that are solely based on shared origin, but because I'm embarrassed to admit I don't know anyone, outside of my family, who is Colombian. In response to the government repression, groups of Colombians in the UK have come together in solidarity with the protesters, lobbying the government and raising awareness about disappeared activists. Speaking to one of them, in Edinburgh, I ask her about her own conversations with family back home. 'It does usually end with that,' she says. 'You know, the *you're not here, you don't know what it's like.*' But she lived there until her early twenties, which suggests that it's the leaving itself that cuts us off, more than the time spent away. Meanwhile, in connecting with others who worry about their country from afar, and feel responsible for it, she says she's found a place from which to act. It's a relief.

We don't know what it's like to be there. The distance creates a longing and a different, wilful belonging, one which may have its own part to play when those most affected by, and least responsible for, a global crisis are also the least heard. My connection to this place — the country it became, and is, in spite of itself — is a narrow stretch, liable to flooding.

It's only a matter of time,
and not only time that mattered.

Nociception, known as pain reception, is really a response to violence. What happens when we identify pain, but not the harm that caused it? The words but not the mouth. What sting, what black and blues over what eyelids and what fists hole up

in those gaps? By the time it happens, what do we call all this —

that's been happening all along?

BIRTH STRIKE: A STORY IN ARGUMENTS

An Argument Against Fear

In an interview on UK morning TV in March 2019, two women in their twenties and early thirties were discussing babies. I was supposed to be doing something, hooking one productive hour onto another, but the babies caught my eye for the reason that they would never exist. The sound of them would never crawl out of a hypothesis. One of the women had founded a campaign group, whose aim was to demand urgent governmental action on the climate crisis, including a decarbonisation of the economy, but also to open a space of solidarity with others who were too afraid to become parents due to environmental collapse. 'You don't want to pass that fear on,' the interviewer offered, nodding, as if this particular fear was a genetic condition, lurking in the blood. 'I'm concerned there is no future,' the woman, Blythe, said in answer, which should put an end to any condition, blood-carried or otherwise.

The most recent climate models demonstrate that we're heading for 5 °C of warming by the end of the century, depending on the roles of aerosols, cloud cover, feedback loops. If I had you now, you'd be over eighty years old by 2100.

The strip at the bottom of the screen read: 'The women not having kids because of climate change'. It didn't mention that they were *choosing* not to bear children. The existence of a choice appeared to go without saying, as if the climate crisis couldn't force childlessness upon anyone, or make us do anything that we don't want to do.

Over a cup of coffee and on an average morning, laughing at your father's improvised songs and barefoot filth covering the carpet, eighty years seems ancient. What promised an autumnal lid of a day with thick rain on the inside has lifted to reveal blood red, pumpkin orange, a river green — the possibility of change in all directions. Eighty years carry an uncertain, therefore endless, number of our weirdo, theatrical mornings.

As if what the climate models said was: we guarantee that your child will live till eighty. They are rather saying the opposite — that I'd never be able to tell you, honestly: 'don't worry, you'll live until you're very old.'[1]

Sitting at our kitchen table, chomping on oat cakes, this 'not having kids' acted as a lifeline, making it deceivingly straightforward to grab, to hold on to. I kept chomping, one bite elation, one bite 'what the hell is this new thing now?', whilst immediately looking up the Birth Strike website. I sent them an email with the subject line 'Others who feel like this'. Next, I texted Adam who was at work, equal parts bored and terrified for quite a few months now. Rude customers never give you the impression that they know about existential threats; the need for impeccable service overshadows

1 As opposed to the main text of this essay, the italicised sections are based on notes I made at the time when Adam and I joined Birth Strike. They also include excerpts from a sound piece called 'Brain Child', which we created together for the event 'Is There a Future?', in London, July 2019.

any sharing of vulnerability. Lately, more young children had started coming in to the shop where he worked, which was unusual, considering it was a whisky shop. He described them as 'flirty', by which he meant that they tried to engage him in conversation as he shelved Lagavulin. Their parents would look over to him and say things like 'oh, she likes you!' — always with a thicker-than-usual Northern accent when he was relaying the scene. My theory was that encounters with small humans took him back to Bradford. He didn't know what to do with their willingness to acknowledge him as a fellow inhabitant of earth, and his subsequent reaction: something like joy. It seemed so perfectly possible to have that kind of life, to parent someone into the future, just not for us and in this late segment of history.

Private and Public Arguments

'To put it in perspective,' says one of the world's leading climate scientists, 'how many of us would choose to buckle up our grandchildren in an airplane seat if we knew there was as much as a 1 in 20 chance of the plane crashing? With climate change that can pose existential threats, we have already put them in that plane.'

You are hungry, and way too hot. I blow my fevered breath on the space between your eyebrows. You learn to walk and there's either a scorched earth or only water under your feet. I find myself picking you up all the time, restricting your contact with our uncertain ground. The aeroplane image makes me realise I'm very unspecific when I hypothesise about injuries to your body — that the future, in spite of so many scientific articles, remains vertiginous, as the suffering of others always does. It's as difficult to imagine what dying in a plane crash looks like. The gap between the knowing and the ground.

If I do not have you, if you do not exist, then something we know for sure is that you won't go anywhere near that plane. The problem persists for every other child, but at least I save you by not having you. Your suffering is possible, as anyone's is, only if preceded by your being born. This kind of certainty counteracts the very nature of futures.

I tell Blythe — wondrous musician, and nowadays good friend — that I'm writing this piece. I explain that I want to better understand what our versions of Birth Strike were, what happened to them, and that I need to ask her a few questions to fill in gaps. We have a long Zoom call and laugh at how I appear to be interviewing her now, after all the times we were interviewed by others, about our wombs mostly — about what we would and wouldn't do with them. There were a few months when I imagined myself walking around with a picture of my empty womb as a necklace, a kind of scarlet letter I'd hung there myself, which was never the same letter to any two people reading it.

I'm asking Blythe the questions I should already know the answers to (what made you go public with such a private choice? Did you feel alone when you first started? Were you worried about being judged?), but it is precisely because I think I know that the questions are necessary — so much of this story is about assumptions. Just because Blythe and I share in each other's stories doesn't mean that our stories are the same.

'Birth is something most people think about at one point or another in our lives,' she says, 'whether we become parents, give birth, or not.'

Birth Strike was a way of telling a story that would affect people around her, and make them listen in a way which scientific news often can't. In her experience, the decision to not have children due to fear of climate collapse was still 'a bit hush-hush' around the time she started Birth Strike, even in activist circles, and it certainly wasn't talked about in mainstream media. Making her private decision public was essential for that reason: the way people weren't talking about it was itself a part of the cognitive dissonance around climate collapse in our part of the world, in our section of that part of the world.

Any parent is constantly scared for their child's safety. I'm told this by parents as well as non-parents, and I don't have the resources to draw on for a reliable counter-argument. I don't know how reality might shift if you rode into town.

Here are some things I do know: 5 °C is, of course, a global average. In the pole areas and inland, it could be twice as hot. The last time the earth experienced 4 °C above pre-industrial levels was about ten million years ago. We have never seen a world like this one before. I can't guarantee anything when it comes to this shitstorm you would call life. In theory, no one ever could, for their children, but this, I know.

My first interview as a representative of Birth Strike was meant to be for *The Guardian*. I was told it would be serious and sensitive, that it would avoid sensationalism, so I agreed, then backed out the next day, blaming the timings and a sore throat. Really, there was something wrong with the light. In the mirror, brushing our teeth, I thought Adam and I looked like the kind of people who would make us nervous. After that, we both agreed to appear in a short documentary. 'Are you absolutely sure?' he said. 'I think so,' I said, stomping forward, forward, intent on bum-rushing the fear, and then I backed out of that too. The network wanted us to be filmed with one of our friends' children, at our home perhaps, for added emotional impact. I called to decline, then stood for a while with my back against the front door, as if they might still come knocking.

Something similar happened a few times. One day, we'd decide that this particular magazine, podcast, or website understood this story, and would ask useful questions — the kind that could possibly lead to someone finding their own response. Twenty-four hours later, I'd change my mind because of a particular phrase ('personal loss' or 'rejecting parenthood', 'the carbon footprint of a child') which came jabbing like a strobe light in the face. Adam was always better at holding back, considering how it would impact us before taking that desperate forward step, before revealing more. I think he understood, from the beginning, that the door between private and public can never be closed completely: it leaks and it doesn't fit the door frame. Every time someone opens it, it's still just us in there, with a very unfinished story.

The moment I feel you, the moment your presence shows on me, or the moment I carry you — which one of these moments will make me into what I think of when I think of 'mother'? Which one will give me a different relationship to my body and everything it does?

During the first interview we did go through with, we sat on the floor of our lounge. Both were squeezed into the space between the couch and the coffee table, without moving the coffee table, for a lower point of gravity, or the protection offered by hard surfaces on all sides. Through the speaker of one of our phones, placed on the table, level with my chin, we spoke to a US-based journalist who may not have had breakfast yet; she sounded starved but resolute, very professional, and extremely young. One of her first questions was: 'How did you arrive at this decision?'

So, we were already heading off-piste. Look at us, I felt like saying (although she couldn't see us), do we look like people who have arrived? We're in our thirties and swaddling ourselves with our Gumtree furniture. The engine driving her initial question was malfunctioning. Although we were publicly talking about it, the arguments about kids or no kids were very much ongoing, still en route, the decision made every day, then remade once more:

And your father says to me without (I am the one who can tell and I could never have imagined this before I met him) meaning it:

What if this is all alarmist and we're all going to be fine? What if we're the deluded ones?

How do you arrive if you don't know where you're heading — three, four, or five degrees above pre-industrial levels, civilisation collapse in five, ten years, or twenty — if no one can tell you with certainty how quickly you, or someone else, will get there?

We take turns in answering that question whenever it burrows into us:

Yes, but what do you mean by we? And what do you mean by fine?

Do I have to be prepared for you to be the child who's never been fine under 'normal' circumstances — the heartbroken, lost, and starving, systemically exploited child — in order to honestly become your parent?

Almost two years later, we talk about this too, Blythe and I: if you're the person who publicly says 'I can't have a child unless things change dramatically' and, for whatever reason, at some point you also become the person who changes your mind, without proof of enough dramatic change, what does that say about you as a person? Does it reveal that you were never serious, or that your fear was never anchored enough in reality? What does it say about the danger itself — does it indicate that it's not as immediate anymore, nor as deadly? That it never was? The womb-letter gains weight.

Here's another fear: that if I did change my mind, people would think I was unbelievably selfish — that I would, in fact, be objectively so. It would mean that, all along, I'd been one of those people for whom the urge to carry a life, to house a being and be their house through life, is so overwhelming that it occludes every other person who needs housing — the responsibility we feel for our own, intimate, family's safety greater than any responsibility for other humans who facilitate that safety. Expectations of permanence emerge the moment private decisions are made public. The public realm demands a certainty from our choices and our identities which birth, or the future for that matter, by nature do not allow. Flawed, in flux or not, the moment it becomes public, the story you tell about yourself is captured for posterity, owned by others, anonymous and known. It was pendular and highly unreliable, my commitment to this cause. How could it not be, if birth underpinned it?

The Population Argument

Rewatching the interview that introduced me to Birth Strike, I'm not surprised when numbers are first mentioned. They make an appearance less than half-way through the conversation. More than a question, the population issue arrives as an offhand formality which hardly needs argument, only a quick confirmation: 'You don't want to pass that fear on,' the host says, and then continues: 'or you don't want to continue the impact of the population.' A full stop at the end, instead of a question mark. 'I'm concerned we won't have a life,' Blythe replies. The host reads out a few examples of online comments. 'We would be culled if we were animals,' someone declares, 'instead we are culling everything. We are the threat.'

What does it say about the cause and effect of climate collapse that the carbon footprint of a child is hauled in instantly wherever climate and birth are brought up? It's always at arm's reach, ripe and ready to start throwing around at people, and it stinks. Before anyone talks about an uninhabitable climate that already cuts children's lives short, they turn the other way: to the effects of a child on the climate. What does it say about which children are being referred to?

In the same present tense as your father is pointing out the gunk in my eyes, it is happening. What comes for us all between 2019 and 2100 is a trail that sinks away and swings out of touch. It's impossible to see it without getting so high up, or so close to the ground, that you lose sight of your starting point. One of your parents gets a lot of gunk in her eyes, and the other often looks at her and says, before coming anywhere near: 'pal, you have eye matter!'

The carbon footprint of a child will (and writing this, my stomach churns at how obvious this is, that it ever needs repeating) of course depend on who the child is: where they are born, when and to whom, what advantages in life will be theirs. After coming up against the 'population issue', again and again, I distilled my stance on it to three main arguments, each connected to the other in ways which the numbers themselves, at a glance, are simply unable to convey.

One: it's not about how many we are, but how we live. I often quoted a 2019 study by Oxfam about how half of the world's lifestyle carbon emissions (that is: the child's food, the child's clothes, the child's transport, and everything else this proverbial child, the one with the footprint, needs and consumes) are created by the wealthiest 10 per cent of the world's population.

Two: following the current trajectory of global emissions, and of heating, controlling population numbers is a distraction we don't have time for. As a strategy for mitigating catastrophe, having 'fewer kids' doesn't even live on the same scale of urgency. The window in which to make a difference is far too narrow for this to be worth discussing.

Arguments three, zero, and four at once: populationist rhetoric, racism, and misogyny have gone hand in hand for centuries, and still do. Many people connect population control with distant pasts, or authoritarian regimes, which is itself nothing but a mark of privilege: a sign that you are one of the wanted ones. A few months before I began writing this piece, reports emerged of women who had been given hysterectomies against their will while in a US detention centre. Population as a focus is not merely a distraction from the much more complicated work so urgently needed — how about dismantling an economic model based on exploitation, extraction, and endless growth? — but a deadly one in and of itself. It provides an excuse for bolstering the same structures of extraction and exploitation, using the racism at their core as yet more building material.

In awkward or outwardly friendly, but hideous, civilised, and painful arguments about population, these have become my three tools. They look so disappointing in my hands, never good enough on their own, in the way that nothing which obscures the infinite complexity of systems, instead of offering a taste of what it's like to live in and of them, is good enough. The numbering is necessary because that is how arguments work: you arrange them into a line of thought, a progression of logic. In reality, inequality informs who gets a choice over their own body; patriarchy, racism, and heteronormativity inform that inequality, and the mess we're in — happening not in the future but between every one of these lines — was caused by all of the above. The arguments are entangled, reacting to each other in webs.

In October 2019, figures from the Adoption and Special Guardianship Leadership Board show that there are twice as many children up for adoption as there are available families in England. You could be one of these small humans. That could be my response. At present, your father and I dedicate an average of ten to twenty hours a week to climate activism, on top of our full-time jobs. If I were to mother you, biologically or otherwise, the flood gates would still open and close at the same time. Survival would be about you, first and foremost, for whom I'd be entirely responsible, my nerves wrapped around you. I think I'd love it. Sometimes, I'd hate it. I tend to get obsessed by people.

Where am I most needed? To whom else do I owe my response?

With such liability at its core, the membership of Birth Strike was unsteady, more jelly than apple, now an outburst of communication, then periods of silence. Very soon, Blythe and I were handling it alone, answering media enquiries and calling on others when needed. In Birth Strike's declaration, we explained that we didn't see population as a core drive of climate change. We made clear that we, in fact, weren't telling people whether they should or shouldn't have children — only that there's a threat so ghastly, and so ignored by those in power, that we're too scared to have any ourselves. In spite of this, the population argument was there, waiting for us in every conversation; every pitched article and journalist request carried with it a whiff of 'there's too many of us'. It was the headline we were allocated by default, after we'd explained to the journalist why that headline shouldn't be used. This wasn't only the case with the media. Having read and subscribed to the declaration, an increasing number of members of the Birth Strike Facebook group went on to share articles and videos about the effects of population numbers on the non-human world. In response, we introduced guidelines, asking members to refrain from posting about population, and from expressing judgement about parenthood either way.

For about a year, my Facebook notifications were filled with signals from the group, this wee nook of the internet. Seeing them flash made me brace, because more often than not, I knew roughly what was waiting. The articles, the videos, and the posts about 'humans' as a non-specific entity which is destroying everything else never stopped, until the group itself had ceased to exist.

Arguments as Judgements

There was this guy one day who posted a video in the Facebook group. He was sitting on a beach, palm trees in the background, talking about his fear of climate collapse. He gave specific examples of what is likely to happen to ecosystems, weather patterns, and human societies in the coming years, and raged at how negligible the action of any government is in comparison to the need. As his anger escalated, he also turned prescriptive: 'Don't have kids,' he said, looking at the rest of us, at anyone who was watching the video. In the accompanying post, he wrote about how grateful he was to finally have found what he saw as a group of like-minded people. 'Don't have kids,' he repeated, and I thought: 'what would you say to me if I told you that the only reason this is hard is because I have this wanting?'

I'm not saying that I don't think people should have kids.

I'm saying that I feel like I can't, although maybe physically I could (by this point, who knows?) and that certainly I'm very, very scared to. This doesn't feel like a choice — a rejection or an acceptance of you, whoever you may be. A choice would be 'I want to' or 'I don't want to', not 'I want to but the wankers who started funding climate denial decades before I was born have made it too dangerous for you to exist'.

I removed the video and posted an upbeat message to the group, asking members to 'please read the guidelines before posting'. The person who had posted the video responded by telling us we were a joke. I took the conversation to a private message, where I wrote that I hadn't wanted to shut him down. I explained why I thought personal judgement wasn't helpful (history of coercion, the human right of ultimately deciding over your own body), and then I offered to talk with him more about our reasoning. A minute later, the response came, in which he told me to fuck off. I really didn't get it; I was just like any other climate denialist, in his view, with a rehearsed line where my spine should be. I wasn't being consistent and my heart, obviously, wasn't in it, otherwise how else could I censor him? At this point, I blocked him, and then I did feel like I had failed, not with what I said, but by keeping anything else from being said. I'd made a choice to talk about this stuff publicly. This is what I should be prepared for.

Looking at that exchange now, it reminds me of the time Blythe went on Fox News and was told by the presenter Tucker Carlson that she 'should have some children'. 'They put things into perspective,' he said, as he glanced at his notes, this white man with a gargantuan platform, wrapping things up. For or against procreation, you're still telling someone what they should do with their body, how they should survive in this world, and survive their body in this world. All such judgements are linked — homophobia and transphobia, as well as misogyny, are siblings in a patriarchal society. Forced sterilisation and forced birth, like conversion therapy, are acts of abuse, and they are advanced by a culture of judgement. If I say 'I am too afraid to have children in this climate', what someone might hear is 'this climate is too dangerous for children' or even 'who could possibly have children knowing this is what the climate is like?' The distortion isn't a built-in mechanism — a natural instinct which as humans we can't escape — it's privilege defending itself, holding on to its power by exerting control over our bodies.

What kind of 'mine' could you be? What kind of 'yours' could I? Long before 'demisexual' was a term, released, juggled, and offered to me by new youth, and long before I met your father, I knew that my wanting never started with sex, nor with gender. Sexual attraction was never wanting's catalyst, nor what sustains my most intense longing for another human. How does this queerness of mine, the wanting that grows very slowly, in the mind, connecting only under its own emotional terms with a kind of sexual desire, define the ideas I have of myself as 'mother'? Or as someone conceiving through sex, carrying a child as part of an outwardly heteronormative relationship, giving birth?

I say to your other parent: 'I remember sitting in the car with my dad once. I was definitely a teenager still, and I told him I didn't want children. He said, "just wait ten years." I think I said it because it's easier to say you don't want something than to say that it is beyond you, that your body simply doesn't work that way.'

Arguments About Choice

More recently, I took part in a discussion on a radio show. Four of us had been invited onto the panel, live-streamed digitally from wherever we were, which meant that I was, once again, at my kitchen table, in the same chair from which I'd sent my first message to Blythe. The host began the segment by addressing me. I had 'chosen not to have children for the sake of the planet', she said. Could I elaborate on this?

There's such a gap between how that question is phrased and any reply I could honestly offer that I had to retreat and adjust her question first. Otherwise, everything would have become a farce. 'Well, first of all,' I started, 'I wouldn't say I'm not having children for the sake of the planet, but for the sake of my child.' I added that the nature of birth and decisions around it are very uncertain (back and forth, a pendulum hanging around my neck), so that, in essence, I actually remained undecided. There was a pause at that point; this was not the story they'd asked for. I waited for her to comment, or to ask a follow-up question, but instead she turned to one of the other speakers. For the next fifteen minutes, I sat and peered out of my window, where the river that day looked like dishwater, and pretended I was long-distance running on my chair to get rid of excess adrenaline, until I realised that they might not ask me anything else, that my role in this discussion was, most likely, played out.

'If we had a child,' Adam says, 'how would we keep them from becoming an asshole? Genuinely, what if they choose to become a banker? Or a Tory?'

You will not be someone who's had. And you will be had in so many ways I can't handle.

In her book *On Infertile Ground*, gender and climate-change scholar Jade S. Sasser explores the role of women as 'sexual stewards'. People with wombs carry the responsibility of perpetuating human life, but not haphazardly. We are supposed to hold our future in perfect balance between our ribs and pelvic floors, to handle it rationally, have kids at the reasonable rate, regardless of where and under what circumstances. Discussions about population, nowadays, are increasingly framed as being all about female empowerment, educating girls in developing countries, and granting them the ability to choose. This sidesteps the racist history of explicit population-control rhetoric, but the empowered girls, Sasser argues, are still based on the caricature of an entirely independent individual. 'The private consumer choice model is so pervasive as part of American culture,' she writes (and I would extend that to 'late-capitalist culture'), 'that it deeply informs our ideas about morality, individualism, and personal responsibility in a range of ways — including how we think about reproduction and environmentalism.'

It's not unlike the food industry's response to broader climate awareness, how hope for collective survival is supposed to be speared on your fork. It's a bit like so-called 'flight-shaming' (which in its original Swedish incarnation wasn't something you made others feel but something you might feel yourself: flight-shame, interestingly, not flight-*shaming*). In narratives surrounding consumerism, survival comes purely from the choices we make in the shopping aisles, and our wombs appear to also have ended up in the shopping aisles. *You have chosen not to have children for the sake of the planet, the women not having kids because of climate change* implies that your individual power is greater than any environment your survival, and your child's, depend on. It's the narrative framework we step into, whenever we talk about birth and climate.

'If we had a child,' your father says, 'what would be the best way to make sure they were interested in music? Should I pretend I was never a drummer and hide all the instruments in the flat? You know that kids are never interested in what their parents like.'

'If we had a child,' I say. 'We could do that. We'd be allowed.'

'What does that mean, to "have" a child?'

How may I have you in a way that's not infected with the ownership that has us all?

Toward the end of that radio panel, I did get asked one more question. It was about personal loss — how I felt, in other words, about my choice. But now I was tired of how I felt, and tired of talking about it. How I felt was the reason I joined the group and started talking about my procreation choices publicly, but talking wasn't supposed to stop at my feelings — it was always about raising the alarm, and bridging the gap between my feelings and the effect they have out there. To me, there's no point in sharing the personal unless it speaks to a broader truth, unless we end up somewhere a little bit new, collectively.

So, I didn't answer. Instead, I talked about what we had intended with Birth Strike and how it didn't go, how judgement snuck in, what happened to the group and why. I wondered if the programme had invited me to tick the box of 'personal' as opposed to the other panellists (a climate scientist, a philosopher, a population-degrowth campaigner) who were there as experts, informed on the social, historical, and political connotations of birth. It might not have been a conscious intention, but we were there to do different jobs, the private and the political kept separate. This mattered. It mattered that the other three panellists, to any public knowledge, were cis men. Although I couldn't see them, I had looked them up beforehand and knew them all to be white. Two of them were firmly against any kind of population control, but it still mattered hugely that they were asked to speak politically about birth, whereas I, as a woman, was the one asked to speak about loss. Even if his argument was aimed broadly, it also mattered that I, a queer woman of colour, was told by the guy on Facebook (who was cis and white) that it's inherently wrong for me to procreate. None of this is solely personal, nor solely political, and the more affected we are by oppression, the more entangled the two become.

I've been looking into ways we could find you. It looks as though most councils and adoption agencies want there to be a room for the child, and we live in a one-bedroom flat. Your father has a criminal record due to breaking rules for what a certain policing entity, on a certain day, deemed was 'lawful protest'. In Sweden, if that's where you are, it becomes very difficult to adopt after the age of forty-two and we're now in our mid-thirties.

Even if you're out there, they may not let us meet. I imagine you riding in on a fox. I suspect it comes from an animated series I was into as a child. It was Dutch and dubbed to Spanish, about gnomes.

I think about the 'sexual stewards' in Sasser's book, those women who are deemed fit to carry humanity forward. These desirable wombs, housing desirable babies, never belong to disabled people, nor to migrants, and they don't belong to trans men. I've also never felt like they represented me, and my intimacies, the way I inhabit my body. In this entanglement of expectations, bigotry, and power, choices are shaped. We never make decisions in airless rooms, with no one looking in and having opinions about what we do, just like we never make them against a blank past.

'I feel like you start a lot of these conversations,' one of your parents says, 'and I'm sorry for that. I'm sorry that I can't seem to bring it up. I think it's just too sad and I keep avoiding it.'

An Argument for Arguments

My mother, your abuelita (or you might have called her something entirely different), will never know you, having died earlier in the year in which I first came across the quote about the aeroplane crashing. Those are the parameters in which I miss both of you. Sometimes, it feels like standing in a house with no floor and no ceiling, not to suggest that you should carry us. Although I'm sorry you'd have to, inevitably, carry our weight.

Either way, I didn't decide not to have kids because I lost my mother. I don't think I want a child because she's gone either. I didn't decide not to have kids because I lost hope. None of this is the end of it. It is the ending where something else takes over.

In September 2020, Blythe and I wrote a statement about the end of Birth Strike. By now, we were the ones left with the decision. Initially, we shared it with our mailing list, and with the closed Facebook group. A few members supported the move. Someone admitted that they'd felt increasingly uneasy about the anti-natal comments and hadn't wanted to engage more because of them. Others were outraged. 'You cannot force us to give birth or have children,' someone wrote to us in an email, which I, honestly, found hilarious. A few people announced that they would simply restart the group on their own, 'someone else can just be admin,' a member wrote, 'DM me?'

'But actually,' I say, 'if they don't know any different, it would just be life for them, wouldn't it? Do kids ever really regret being born?'

Recently, I spoke to a good friend of ours, a single mother with two children in their pre-teens, who's also more clear-eyed about climate predictions and the likelihood of collapse than anyone else I know. I wanted to know if she would still do it. She loves her children, and wouldn't go anywhere near the thought of wishing them un-born, but would she make the same decision if she were faced with it today?

I somehow expected the answer to be tied to the love for them — for you — for the child, once you've met it, to eradicate any doubt. Instead, she said she didn't know, and that she understood that this is really, really hard.

I ask Blythe if she ever thinks that shutting down Birth Strike was the wrong thing to do. There's something like shame lurking backstage on my part. Very early on, she says, after the first wave of media interest, people asked what was next for the 'movement'. She found it a strange question, because Birth Strike was never a solution, or a strategy. 'We never set out to recruit,' she says. To both of us, that would always have been a horrible thing to do. Our group never suggested a tool for tackling the climate crisis, other than telling one of many necessary stories of what we are losing, what so many people have lost, and refusing to lose it quietly.

This also made the community extremely vulnerable. By focusing on the alarm, we failed to bring attention to the kind of action needed, and by doing so at a particular moment in time, when there was a wider public awakening to the climate crisis in the Global North, we opened ourselves to co-option. Stories are the beginning of arguments, and arguments in themselves. People grabbed on to this one and tugged in different directions; a mesh of hundreds of anxieties and needs, its own system of fired-up neurons. Some saw in it the possibility of openly rejecting parenthood in a society which assumes it as the ideal. Some were there because of their deep allegiance with non-human beings;

for them, Birth Strike was a way of siding with other life forms. All of these people claimed the space as their own, but they were never in it on their own. I never intended to be part of this argument, but my, or anyone else's, intention was only ever one part of it — one side of a meeting place. The mistake, really, Blythe thinks, was imagining we could control how a story is received, and it's a humbling thing to learn. 'It's naive,' she says, 'to think that you can chuck your voice into a melting pot and expect it to be echoed back at you,' to assume you can control other people's responses.

Instead of offering a straightforward yes or no, our friend said to me that it's my fear I'll have to deal with — my non-negotiable worry for your wellbeing, my anxiety in trying to foresee and to reduce your pain. I know your life would be harder than mine has been. It may also be open, full of connection in ways I have never known. How you bring the world into being around you is beyond me. I don't want it to be beyond us.

A while ago, Jade Sasser asked me where I am now with my decision about children. She's someone else I came into contact with by sending an email, chucking out fishing lines to see who else was there. It was a response that, sometimes, made me less afraid, but this kind of response also opens you up to scrutiny and to having your mind changed. I make my personal, private decisions not in spite of, but because of others, everyone who's knocked on my door, every hurt and good feeling. Every time I've been told 'well done' or 'how could you?', that I'm strange or normal, attractive or a creature of an entirely different sort, every argument I've had and how it's torn me open. I am remade by these arguments. Jade's thinking, through the book she wrote, which both Blythe and I read, influenced my basic assumptions about birth. I told her this. I also told her that, sometimes, it feels more possible to expand our family than it did a year ago. It feels less signed and sealed, less as if it's already gone and done for, as my fertile years leak down the side of our bed, through the windows, and into the river that runs through Bath, as fear moves through us and a massive chunk of the Arctic moves south.

'This couple came into the shop today. I'm not sure which different ethnicities they were, but their kid was this really interesting-looking mash-up of their features and colours, and it made me sad because I'll never know what a mash-up of the two of us would look like.'

'I want to know what you mixed with me would look like,' I say.

'We could just use face-merge software.'

Like I've said to your father at times: I could never have imagined you.

It's not that I feel more optimistic. Rather, I've come to think that in trying to decide whether it was right or wrong to bring a child into this now, I was asking the less helpful questions, the ones less anchored in the now I'm living in. It would also be disingenuous to argue that my ambivalence about motherhood is separate from the kind of mother that society expects me to be, from a lifetime of becoming friends with one's body. The way forward must never lie in whether or not to give birth, but in creating the circumstances in which it is safe to not only give birth, but to create all kinds of families, and for those families to look after each other. The urgency lies in fighting to bring down the structures that judge us for how we survive in our bodies — they are also the systems and attitudes which stand in the way of real transformation and collective survival.

If there is no other now to choose from — how do I best respond to this one? Because there is still uncertainty, there is still this wanting.

'The thing is,' I say to my Person.

'The thing is, I never wanted a child in theory. I want one with you.'

This is all a bit private. 'We're really kind of private people,' Adam said at one point, by which he meant that we generally don't talk about personal things except with close friends. I struggled with this — the worry of being seen as attention-seeking — and in that tug, the very chaotic, radical nature of birth once again shines through. It happens behind closed doors, but it involves all of humanity — it is legislated on and used as a tool of power and ideology, but it is ours, at our most vulnerable. Few things highlight the permeability between private and public as much as birth does. Very few, but this so-called 'end of the world' might be another.

Postscript

The name Birth Strike has been used by other groups and movements. It's an evocative and visceral name, which also highlights the inherent labour in giving birth. It was recently brought to my and Blythe's attention that a group of people have decided to restart Birth Strike in the UK as a form of climate activism. From what I have seen and am aware of, the people involved do not agree with our non-judgemental stance on birth, and the reason we closed down Birth Strike. We communicated our reasoning to them. The arguments continue.

When do we give each other a signal so we can sing it — and the ability to belt it out?

Not to feel responsible but to ably respond.

ON WHETHER OR NOT TO THROW IN WHOSE TOWEL: A PERSONAL ENCYCLOPAEDIA OF HOPE

En memoria de Olga Johannesson

Hope as a Towel: 'Todavía no vamos a colgar la toalla,' my mother used to say after an argument with my father, when my sister and I started crying 'divorce!', like two little outraged villagers. We worried they would, which they never did. 'No cuelgues la toalla,' she said to me, close to midnight the day before a school project was due. Aside from being ill-equipped at managing time, I've never been too sure about when time has run out. 'No voy a colgar la toalla,' she said, later, about her own chronic, then terminal, illness. So many towels about to be thrown in, yet hovering above the floor. You'd think that, in our family, we did a grotesque amount of washing.

Hope as a Toad: The Uruguayan writer Eduardo Galeano described the sound hope makes as that of a 'minuscule toad', 'un sapito minúsculo', in the grass. The extra diminutive in 'sapito', particularly Spanish, makes the toad a tender and jumpy kind of small thing, easily scared away. I found this quote online and have attempted to trace it. Several sources state that Galeano said it to the paper *La*

República in 1993, but I can't find the original interview. It makes me stop to ask: does it exist? Hope as hearsay, unreliable, not only easily frightened but almost impossible to spot in the first place.

Hope with Sides: On the day my mother received her first diagnosis, aged forty-seven, she used touch as an image for hope, although it was such a built-in metaphor that it didn't advertise itself as one. My mother, my father, my sister, and I sat around a table in a hospital garden. I think there were cups of tea in plastic mugs. My mother said that 'nos tocó', meaning that it was our family's turn. We'd been so lucky until now; we'd been hugely privileged, in fact. 'Nos tocó' translates as 'it touched us', but the 'it', whatever it might be, is silent, open to definition from surrounding context. In this case, *it* was a tumour in my mother's cerebellum. Although benign, it posed an immediate threat and had to be removed before it grew further. She'd started to ask me to drive the car that summer, the summer after I turned twenty, in order to teach me, but also because she always felt drunk.

(It) touched us. I thought of kids in a school yard being chosen for sports teams, potentials being tapped on the shoulder, or someone's turn to do the dishes. It touched me, and then it would be someone else being touched next time. Touch is by nature temporary; any permanence makes it something else: a pressure, a holding. Because 'it' touched us, it meant that we could get through it; hope being the idea that 'it', this crisis, had sides and could be constrained.

Hope as Comfort: Someone I used to work with asked me if I really thought there was any hope. We'd never really spoken about the climate crisis before, although I've developed the habit of speaking to most people, at some point, about climate collapse. His boyfriend had started devouring climate predictions, for the first time, during one of the UK's Covid-19 lockdowns, and now it was all he could talk about. 'He's slipping into nihilism,' my friend said. 'I could do it too, but not at the same time as him, obviously. Where would we be then?' We talked about the kind of things I'd been involved with, in the realm of the 'doing something'. I said that doing something tends to help, that it's the lack of a response that makes everything so much worse. 'Do *you* think there's hope though?' he insisted.

Who the hell qualified me? To hold someone else's hope in my hands.

Hope as a Towel #2: Famously, however, the term 'throwing in the towel' doesn't come from washing dishes, or washing anything, really, but from the boxing ring. This could help explain why it travels so easily between languages. Although the towel itself used to be a sponge (still is in French), and although the towel is 'hung up' in Spanish, rather than thrown in, the image is still that of a body, clashing fists, sweating floods, and bleeding on whatever piece of cloth is used in that particular locale, resisting and refusing to leave the arena where its fate is decided. When the cloth is thrown into the ring, or hung up on the side of it, as the case may be, someone has officially given up hope. That is, the hope of winning, you'd have to assume. You'd expect that the meaning of the thrown-in towel has been established beforehand, and agreed upon by all involved.

Hope as One More Load: For eight out of the fifteen hours of my mother's brain surgery, I stared at two industrial-size washing machines. It was a summer job at a bakery where they made biscuits, truffles, a thing called 'dammsugare' — filled with, it was rumoured, crumbs from every conveyor belt in the building and then covered in marzipan — and many, many cakes, to be found in supermarkets around Sweden. Purveyors of fine confectionary to the king, their cube-shaped industrial bakery, where half of the employees were from the immigrant Chinese community and many were mothers and daughters, was a short bike ride from my parents' house. That summer, I was in charge of laundering the staff's dough-encrusted uniforms.

I spent a few hours of that day copying text messages from my phone onto a scrap of paper, then I tried to read *The God of Small Things* by Arundhati Roy, until someone told me not to let management see that I was reading, or they might think I didn't have enough to do. 'I'd like something,' I said, 'to do'.

For the rest of the time, it was me and the tumbling portholes of the machines, the folding of white canvas shirts every forty minutes. There was one hour in which I mopped the floor of the women's changing room, listening to harp music, and then there was one hour less to go. There was the next load, and the heat of the tumble dryer, my mother's brain in suspension under able surgeon's instruments, and my actual mother hovering somewhere above all the nerves and their endings, waiting to come back down, for the crisis to end. No pieces of cloth thrown anywhere, except where they needed to be, in order for me to get paid.

Hope as non-Courage: 'We need courage, not hope,' wrote Kate Marvel, climate scientist at NASA's Goddard Institute for Space Studies, in a 2018 article. She 'used to believe there was hope in science'. The sheer possibility of it, that we can measure temperature, know causes, save lives, cut nerves, and make the body behave differently, make parts of it speak to each other according to our intentions: hope as the door opened by future discovery. That there might be a cure? But, she says later in the piece: 'I have no hope that these changes can be reversed. We are inevitably sending our children to live on an unfamiliar planet.'

This is confusing to me, because in the matter of next breaths, of being able to simply exist in the following morning and not go mad, I always thought hope was the prime suspect. Hope, rather than the opposite of courage, was what made courage possible.

Hope as a Return: There was some haemorrhaging during my mother's surgery. The neurologist was never able to say, with certainty, if this is what caused the complications, what she eventually came to call 'my illness' or simply 'lo mío', my thing, mine. Roughly six months later, when the crisis was supposed to have ended, been overcome, my mother's head began to twitch downward and to the left side. This made it increasingly difficult to stay upright, to hold her balance and to walk. She was diagnosed with cervical dystonia, a condition of the nerves which causes them to send involuntary signals to muscle groups around the spine. The nerves gain a will of their own; the point where they used to connect seamlessly becomes conflicted, an altercation of signals which never stops. The muscles are forced to spasm, regardless of what the person, to whom they belong, intends to do with them.

As a result, my mother often came to feel as if she didn't belong in her own body anymore. She talked about her body not allowing her to live. Her body was what kept her from being herself, by which she meant the person she had been before. She was astonishingly disciplined with physiotherapy, rising each morning at dawn to exercise in whatever way the nerves and confused muscles allowed. She didn't want her body to win. She never missed a beat.

Hope as Courage: In the 2015 introduction to Rebecca Solnit's book *Hope in the Dark*, I find that I'm not alone in mixing my hope with my courage. Hope, Solnit writes, is 'the belief that what we do matters even though how and when it may matter, who and what it may impact, are not things we can know beforehand'. Hope, then, as a starting point for action, which in fact reminds me of Marvel's definition of courage: 'the resolve to do well without the assurance of a happy ending'.

Is a 'happy ending' the same as believing that what you do matters? I think not; rather, it's the 'doing well' and the doing something that 'matters' that seem to speak to the same power, call it hope or courage, depending on how you define one or the other.

Post-Hope: Philosopher Jan Zwicky illustrates humanity's current predicament with the story of the ship of Delos in Plato's dialogues. Socrates has been condemned to death. Everyone knows that when a particular ship is spotted on the horizon, it's time for the hemlock. 'Humans collectively are now in Socrates' position,' Zwicky writes, 'the ship with the black sails has been sighted.' She goes on to discuss how to live under such circumstances, with courage, a sense of justice, self-control, and compassion, rooted in Socrates' thinking. How people live matters, even though it might not make a difference to the ultimate outcome.

That part makes perfect sense to me and I still disagree with Zwicky's central premise: the positioning of those suffering a final verdict. Such is the unjust nature of climate and ecological collapse, mirroring the unjust make-up of society, that I can't think of a single experience of this crisis that could apply to all 'humans collectively'. This kind of generalisation echoes the later part of Kate Marvel's article about hope. 'The sheer scale of the problem,' she writes, 'provides a perverse comfort: we are in this together. The swiftness of the change, its scale and inevitability, binds us into one, broken hearts trapped together under a warming atmosphere.'

I don't believe either of them mean it that way, and yet I want to shout at them: but it *is* easier for you. It's easier for me. It should be obvious by now that climate collapse doesn't eliminate inequality, it is fuelled by it and widens the gap. Either side of these unavoidable chasms, whose hope is being dismissed?

Hope as an Assumption: It's not that I have no idea what people mean by it anymore. It's more that everyone seems to mean something different. Sometimes, they mean different things at different times. The effect on its surroundings, every time the word is used, varies. It falls on variously shaped ears and with varying intentions, yet the word 'hope' is accepted, thrown about, as if it was one, consolidated notion. It's not that I've lost it, but that the word appears to have fallen into a vat of boiling water, lost its skin. It is so utterly everywhere, that it is nowhere to be found.

Hope as Dreaming: Foreign aids and unwanted helpers moved into my parents' home, and stayed for the remaining thirteen years of my mother's life. There was an indoor walker and an outdoor one, an aluminium cane with three legs at its bottom (which always made me think of the way my sister jumped into a pool or lake when she was younger, all streamlined elegance until right before the splash, when she'd pull her legs up and turn into a tarantula — my mother pointed this phenomenon out to me once at a water park), another cane, wooden, referred to as 'bocken', the goat, cushions for holding one's head up, pill organisers, and handles in the shower. At times, it seemed like the helpers, rather than facilitating everyday activity, stood in the way of my mother being herself. They belonged to another self, who had come back down from surgery instead of my mother. The way forward was also the way back, but science didn't suggest this as a possibility.

She told me that when she dreamed, she was never ill. She'd been running last night, or driving a car, as if there were nothing to it.

Hope as Honesty: Something I've spent too much time thinking about (if by too much I mean time that could have been spent either on mutual aid or growing vegetables) is the Dark Mountain Project. A collective of creators, a cultural conversation, and series of books, it was launched in 2009 by writers Paul Kingsnorth and Dougald Hine. A key part of its mission is to hold a space for honest grief in the face of ecological collapse, rejecting narratives of sustainability, activism, and, above all, what they view as humanity's myths of progress. In an interview from 2012, a reproduced email exchange, Paul Kingsnorth explains that whenever he hears the word 'hope', it makes him 'reach for [his] whisky bottle'. 'What does it mean?' he says. 'What are we hoping for? And why are we reduced to something so desperate? Surely we only hope when we are powerless?'

It's a lot, a rollercoaster of a statement. Hope is useless and an empty word. Kingsnorth is not sure what it means, yet he rejects it as a sign of impotence. It's desperate, as desperate as the whisky bottle? More than anything, the sentiment begins to illuminate a presumption about what hope is, shared by many in the Global North environmental movement which, ironically, Kingsnorth left.

Hope as Privilege: Recently I told Zoe, a campaigner for decolonisation of disability rights and member of a collective of grassroots organisers, about my confusion around courage and hope, how I still can't get the juxtaposition in place. We spoke about the slogan that Extinction Rebellion used around the time of their foundation: 'Hope dies, action begins.' At the time, it sounded to me like a last-minute call away from excuses, which is exactly what it was, but for a very specific group of people — those who'd had the privilege of making excuses.

'Hope framed in a very western way,' Zoe said, 'can take the form of moral and political laziness.' This, it seems to me, is the form of hope that's being rejected by Kate Marvel, XR, Paul Kingsnorth to some degree (a desperate hope), and even by Jan Zwicky when she talks about spotting the ship from Delos. It's the basic assumption of safety, regardless of the world's undoing. 'That's what the 1% is,' says Zoe. 'People can afford not to be hopeful because they don't have to be.'

Privilege means trusting you'll ultimately be okay — there will be a last resort. The symbol at the heart of the Dark Mountain Project is particularly telling. Who has access to this mountain (or in the case of Kingsnorth, a farm in Ireland), from where to contemplate and grieve the destruction of everything we've known? It seems to me that the turn away from hope, so prevalent in recent climate activism, is really the rejection of a specific definition of it — one which was never an option for so many. If by hope you meant trusting that things would ultimately be fine, then yes, courage is better, and more honest. That definition of hope, however, was always privileged. If this was your hope, you were always in the minority.

Hope as a Legal Drug: Heart palpitations are a very recent acquaintance. I spend a day reading up on possible causes: 1. Perimenopause 2. Water in my lungs 3. The crises, as a canvas to our days. After paramedics visit our flat one night, at least I know it's not the second. The first thing any sensible person would do is to quit drinking coffee. For the first time ever, I switch to decaf for a few days, and the low tides, a kind of hopelessness, follow closely on the heels of sobriety. I can't seem to jump onto any higher ground, not even a chair. Wow, I think, was my hope always brewable? Something so fickle and liable to being drained.

Hope as Botox: There is no cure for dystonia. What I mean by this is that there is, as of yet, no way to return the body to its former working ways, nor to completely stop the pain. In some cases, surgery is considered, but this was not advisable in my mother's case. A common treatment involves Botox injected into the muscle to counteract the spasms, treating the effect of the disease by injecting nerve poison. Rather than healing the rift between nerve and surrounding tissue, it severs it further, but there is no alternative. It helped, sometimes, but the treatment itself was also very painful, a journey of constant trials and intermittent errors. Sometimes, the neurologist would hit the right muscle, often they wouldn't, and sometimes it affected my mother's ability to swallow and to speak. The body is too complicated for silver bullets.

We used to joke about how Botox is what Hollywood stars use to reduce their wrinkles. My mother was having Botox. We would say this, and we would laugh, because wouldn't that have been the least expected turn of events, for anyone who knew my mother?

Hope as Change: 'As a chronically ill person, for me hope is a lifeline,' Zoe tells me. 'As a Jew, I'm not allowed to not be hopeful.' The idea of hope as 'fixing things forever', Zoe suggests, is also a very Christian one. As a somewhat pagan, mostly agnostic person, I can still see that it's all-pervasive. Any story of hardship in popular western culture has overcoming — succeeding, winning, the body or the earth healing once and for all — as an ideal ending. Other faiths, including Judaism, hold the need for constant repair of the world as a core duty. A hope which relies on ultimate salvation is a shirking of one's responsibility.

This is intimately tied to an idea of progress. It reminds me of the Dark Mountain manifesto, and the forcefulness with which Kingsnorth and Hine reject progress. They seek to meet collapse head on. But isn't collapse itself the flip side of progress? Ancient civilisations are portrayed as gone forever, although it's never that clear-cut, although traces and heritage remain. A crisis is always portrayed as having an ending, which leaves little understanding for how chronic illness works, how it doesn't go away but, like anything, goes through changes. Progress and collapse suit a linear narrative as mutually dependent opposites (either a cure or death, either a getting better or not getting anywhere at all). This makes me wonder why I always use the term 'climate collapse', instead of 'climate change'.

It's because of the harm, I think, which is and has always been an act of violence. It's because this change didn't come from nowhere, from no one.

Hope as What Hope Does: Eco-philosopher Joanna Macy differentiates between active and passive hope. Passive hope, she argues, depends on 'external agencies', whereas active hope 'is about becoming active participants in bringing about what we hope for'. I am drawn to this, not only because it acknowledges that hope is a hydra with many heads, but because it points to how hope *does* things. As well as what you're hoping for, and what that hope feels like, there is also the effect that our narratives of hope have on others, and the responsibility this brings with it. When a middle-class person in Scotland says that there is no hope with regards to the climate crisis, what does that do? When someone with the means to retreat to higher ground gives up, instead of continuing to work for change, what power is being withdrawn and who is being abandoned? The effect isn't just interpersonal, and it doesn't only happen in moments of private fragility. Public attitudes, movements, and their successes are determined by the hope that people instil in them, and if those with platforms and the ability to be heard say that there is no hope, that is an active stance against the action which may save millions. Nihilism, voiced from a position of privilege, is an act of attrition.

Hope on a Staircase: There was this one time, about five or six years into my mother's illness, when I was staying with my parents for a few days and came off a phone call. I went to tell my mother about something, but couldn't find her. Having noticed that the flat door was open, I went out to the building stairwell. She was sitting on the top stair, holding her head up with one hand, as she often did so she could see you properly, speak with you. She was leaning with her upper body against the wall. 'I'm tired of this shit,' she said, in Swedish, as she said on more than one occasion. She never said this in Spanish. We sat there together, talking in a stairwell about how shit it was, and what could possibly make it less so.

Later the same evening, I went to meet up with a friend at a café. I said to her that, sometimes, I wondered if my mother would be happier if she accepted that things would never be the same again, that she would never be cured, that hope could no longer mean hope for a return to a former life. I wondered if this was something that I should say, or at least try to say, to her; if that would be, somehow, helpful. I never said this. I wasn't the one feeling it. It wasn't my hope to give up on.

Hope as a Towel #3: The Cambridge dictionary tells me that to 'throw in the towel' is to 'stop trying to do something because you realize that you cannot succeed'. This makes it sound like a much less complicated affair than a boxing match would be, with fewer stakeholders. In fact, the towel, or sponge, isn't thrown by the person taking the punches, but by their coach. If you're exhausted and bleeding from every pummel, it's a bit too much to also divert your attention to towels. How would you even get a chance to reach one after an upper cut?

What also stands out for me is the word 'succeed'. This was not the way it had to be, or the way we need to look at ourselves. The very idea of life as a boxing match brought us here, and it desperately needs repairing.

Hope as Change: Recently, Zoe tells me, they've begun to express more clearly what they'd like people to say to them when sharing news about their health, and what kind of attempts at comfort are less welcome. 'It'll get better' is at the top of the list of unhelpful comments. The facts underpinning Zoe's illness make such hope hollow. Hope, for them, is rather to fight for a society that holds disabled and chronically ill people better, that speaks with all of us and listens to all our needs.

Hope as Solidarity: I never did tell my mother that there was no hope of going back to the way things were, perhaps because I suspect she knew it, but needed all these different levels of hope in order to find courage. It wasn't my job to remind her of the things she had no power over, which were already done, but to encourage the power she did have to grow, and to stand with her, to oppose the structures that made her life difficult and to celebrate her own reparations of self and world, from the way she advocated for fellow disabled people to her rare ability to nourish others' dreams. Although it is tempting to simplify hopes into 'false', 'honest', and otherwise, we do this from specific positions. I wasn't responsible for my mother's pain, but for how I responded to it. Instead of telling her to be realistic, I'd tell her that she would feel different, because change is inevitable, and the end is the bridge to something else. That we'd do it together. Really, I was saying: don't leave.

Hope as Comfort #2: Adam and I lose hope at irregular intervals, and always in new ways, under a different piece of furniture, and thankfully not, usually, at the same time. I think, perhaps, that this is not about luck but taking it in turns; in other words, it's about responsibility, that which we have for each other.

Adam reads a new article on deep adaptation. 'No kind of action seems over the top anymore,' he says, ribcage looking so open to attack, hands empty toward a cast-iron crow we call Erskine. 'Nothing seems like too much now.' I know that he's right and I know why. I know that whatever is done now, however many subsequent miracles happen and however much emissions are reduced tomorrow, which they are extremely unlikely to, it will continue to get hotter, harsher, less habitable. I can't say that it's not that bad or, the cheap drug of comforts, that it 'will be okay'. But, and this is how unfairly hope is distributed: this happens to be a better day for me. Hell knows why. Perhaps it's because I've had just the right amount of coffee, or because I didn't read that article. Perhaps the coffee, the good day, were the reasons I didn't read the article, and it means that, although I know that all he's saying is true, I can keep us going this time. I have it in me to tell him that everything we do might make it less shit, you know, for someone.

Hope as Specificity: With this in mind, whenever I am tempted to say that it's too late to respond, I force myself to be specific. Don't be lazy, I think. If you're going to give up, then at least tell me what you're giving up on. Too late for what? Is it too late to save millions of lives by halting fossil-fuel extraction now and averting the very worst effects of global heating? Is it too late for opening our borders to people needing new homes, and taking responsibility for our countries' part in setting off destruction? Is it too late for land reforms? Local food security? Or do I mean that it's too late for me, personally, rather than someone else? Am I okay with that?

Specificity challenges easy answers and asks of me to step up. Staying in the space between denial and nihilism, to have that kind of hope, demands courage, because both extremes are easy, and neither hold the uncertainty of survival, of life.

Hope as Dreaming #2: I very rarely dream about my mother since she died. When I do, I think: there's hope for me yet.

Hope as Being Special: Kent, a Swedish indie band adored by, it seemed, everyone but me in my early 2000s teens, had a line of lyrics: 'du är hoppet i ett IQ-fritt land' (you're the hope in an IQ-free country). I never liked the band much, but who could reject that line?

Hope as a Towel #4: Who won? This is a terrible question. Try again: did she get to keep fighting?

Hope as Code for Hope: Today there are 14,423 quotes tagged with 'hope' on Goodreads. Scrolling through them makes me feel like I'm watching ads. I do it because only some of them mention the word 'hope' at all, but they still claim it. There's Emily Dickinson with her explicit thing and its feathers, but sometimes hope is an implicit horizon (Faulkner), other times it is stars (Martin Luther King). Hope is assumed, and feels somehow endangered — washed out in the mainstream. As with Galeano's toad, these are descriptions of what hope feels like, regardless of what is hoped for, or whether there is a scientific basis for hoping. We go: hey, I know that feeling, and we sit on it for a while, let it seep through. It's hope universal, hope as something like faith or being in love.

Hope as a Human Right: The difference being that people don't, very often, say: there's no love left. If someone told you that love had died, you'd have every right to turn around and say: 'hey, speak for yourself.'

Hope as a Human Right #2/ Hope as Dreaming #3: My mother introduced me to the work of Eduardo Galeano, as she did with so many Latin American writers. She wanted to future-proof my Spanish, I suspect. In 1996, Galeano wrote in the newspaper *El País* that 'the right to dream does not figure in the 30 human rights' of the UN. 'But if it were not for this,' he continues, 'or for the waters it gives us to drink, other rights would die of thirst'.

He doesn't say 'hope' either, and as I try to advocate specificity, or have been, so far, I'm not sure this one even qualifies. But what the hell, clearly we're in the same general business. Inching and inching toward something better, for someone who needs it the most.

Do certain questions have it in them

to open the flesh to others' blood?

tick, tock, lightning and shock

to other roots and branches?

Why the nerves?

Because they wreaked havoc with her body
but she was never a nervous person

whereas I: ett nervknippe.

Because there are hurricanes in the brain, as tangled a climate
as we'll know — because what is the Gulf Stream if not a nerve

to be kept alive, alive, to please be kept wide open?

ACKNOWLEDGEMENTS

I write, always, in response to others, being, like all of us, a result of those around me. This book is, in part, a map of connections, irrevocably tied to everyone it's been lucky enough to come into contact with, and everyone I've learned from. I'm forever grateful for the questions you gift me, and the easy answers you withhold, at a time when easy answers are exceedingly seductive.

Thank you to the editors of *Wasafiri* magazine for trusting the odd pairing in '"What Have I Done?" and other Illusions of Control' enough to give the piece a first home. Thank you also to Danny Denton for opening a space in *The Stinging Fly* for 'Freak Aguacero'. Elements of 'Mixed Signals: Five Moments of Un-Belonging' first appeared in 'Thirteen Ways of Looking at a Crisis', a short piece published as part of the Aitken Alexander Isolation Series on the Aitken Alexander website in the spring of 2020. A short section of 'On Whether or Not to Throw In Whose Towel' appeared in a blog post entitled 'Always Ask: Too Late for What? Thoughts on the Day of an IPCC Report' on the Lighthouse bookshop website.

The completion of this book was supported by an Authors' Foundation Grant from the Society of Authors. I'm immensely thankful for the time and encouragement this grant brought with it.

Thank you to the splendid team at Scribe, in particular to my editor Molly Slight for pushing my questions toward greater honesty and clarity. Thank you to Adam Howard for championing the book out there and to David Golding for making every sentence better. Thank you to my agent, Lisa Baker, for helping me carry these essays from the beginning, with empathy, wisdom, and much patience.

Some wonderful humans, their thoughts, their activism, and the communities we've shared, are woven into this book, explicitly and in the gaps. Thank you to Doug Teeling, Paul Reid-Bowen, Peter Birchenough, Iacob Bacian, Euan Monaghan, and Rosie Jones. Thank you to Finlay Asher, Todd Smith, 'Chris', Blythe Pepino, Jade S. Sasser, Zoe Bouhassira Miniconi, and my pals in Global Justice Bloc for sharing your crisis responses with me. Thank you to Nicholas Herrmann for reading the first few essays and telling me they could, indeed, be called essays. Thank you to Annie Rutherford for your expert eye on the 'nerves'. Thank you to Rosie and Kim Sherwood for making and giving the notebook used. Thank you to my former and present bookshop families at Mr B's Emporium in Bath and the Lighthouse in Edinburgh.

These essays were sound-tracked by Explosions in the Sky, Zoë Keating, Loma, and Loreena McKennitt. Although they have no idea, their music also helped with the questions. Thank you, so very much, to every writer quoted in these pages.

Tack Elix, life-sibling, för insikter det tog mig femton år att hinna ikapp. Gracias Tita, Tavito, Martica y Sali por ayudarme a recordar. Tack Jen, systra mi, för att du läste och sa att det gick bra att skriva om oss. Tack Pa, för att vi kan ge oss in i frågorna och ut ur dem igen. Gracias, Ma, sin ti ningún libro. Thank you, Adam, for being my Person. To quote The Decemberists, this is why we fight.

Thank you, reader, for staying open.

TEXTUAL SYNAPSES: NOTES AND WORKS CITED

The following list of comments and works cited is intended, foremost, as an acknowledgement of the network of texts that created this book. Without these connections, no nerves, and very little learned. With that in mind, I have retained the non-English-language sources that helped me write. It's my hope that English-language readers will still find them helpful, and seek out connections of their own.

'WHAT HAVE I DONE?' AND OTHER ILLUSIONS OF CONTROL

p. 4: The article in question was: 'Major Climate Report Describes a Strong Risk of Crisis as Early as 2040' by Coral Davenport. *The New York Times* 7 Oct 2018. Accessed 28 Dec 2021 https://www.nytimes.com/2018/10/07/climate/ipcc-climate-report-2040.html

p. 4: An example of these headlines was in *The Guardian* on 8 Oct 2018: 'We have 12 years to limit climate change catastrophe, warns UN'. Accessed 28 Dec 2021 https://www.theguardian.com/environment/2018/oct/08/global-warming-must-not-exceed-15c-warns-landmark-un-report These headlines would later be criticised for being alarmist, and for underplaying the urgency for many areas around the world. This points to the audience these headlines were aimed at.

p. 7: Peter Kalmus describes his climate terror in the article 'The Climate Crisis is Worse Than You Can Imagine. Here's What Happens If You Try' by Elizabeth Weil. *ProPublica* 25 Jan 2021. Accessed 28 Dec 2021 https://www.propublica.org/article/the-climate-crisis-is-worse-than-you-can-imagine-heres-what-happens-if-you-try

p. 8: The Audre Lorde quote comes from the essay 'Poetry Is Not a Luxury', found in *Your Silence Will Not Protect You*, London: Silver Press, 2017. It was first published in *Chrysalis: A Magazine of Female Culture* No. 3, 1977.

p.8: Eliane Brum writes about the Amazon and the climate crisis here, amongst other places: 'In Bolsonaro's burning Brazilian Amazon, all our futures are being consumed'. *The Guardian* 23 Aug 2019. Accessed 2 Jan 2022: https://www.theguardian.com/commentisfree/2019/aug/23/amazon-rainforest-fires-deforestation-jair-bolsonaro

p. 9: I began to explore thinking around slow violence thanks to *Slow Violence and the Environmentalism of the Poor* by Rob Nixon. Cambridge, MA: Harvard University Press, 2011.

p. 12: The Lauren Berlant quote comes from *Cruel Optimism*. Durham, NC: Duke University Press, 2011.

p. 13: The idea of environmental neurosis is discussed by Johannes Lehtonen and Jukka Välimäki in the article 'The Environmental Neurosis of Modern Man: The Illusion of Autonomy and the Real Dependence Denied'. Although it doesn't discuss eating disorders, it was hugely illuminating in writing this essay, and re-assessing my relationship with my body throughout my illness in retrospect. It is published in the anthology *Engaging with Climate Change: Psychoanalytic and Interdisciplinary Perspectives*. Ed. Sally Weintrobe. New York: Routledge, 2012.

p. 14: 'What If We Stopped Pretending?' by Jonathan Franzen. *The New Yorker* 8 Sep 2019. Accessed 30 Dec 2021 https://www.newyorker.com/culture/cultural-comment/what-if-we-stopped-pretending

p. 14: '"We did it to ourselves": scientist says intrusion into nature led to pandemic' by Phoebe Weston. *The Guardian* 25 Apr 2020. Accessed 30 Dec 2021 https://www.theguardian.com/world/2020/apr/25/ourselves-scientist-says-human-intrusion-nature-pandemic-aoe

p. 15: 'A Succinct Account of My View on Individual and Collective Action' by Kevin Anderson. 24 Aug 2016. Accessed 30 Dec 2021 http://kevinanderson.info/blog/a-succinct-account-of-my-view-on-individual-and-collective-action/

p. 17: bell hooks' words about solidarity come from *Feminist Theory: From Margin to Centre*. London: Pluto Press, 2000.

THE WAYS WE USED TO TRAVEL

p. 27: Mark Augé writes about non-places in *Non-Places: Introduction to an Anthropology of Supermodernity*. London: Verso, 1995.

p. 30: A study published in November 2020 suggests that in 2018, 11 per cent of the world's population travelled by air. For international flights, the figure was between 2 per cent and 4 per cent, further highlighting the nature of flight as a luxury. Findings are published in 'The Global Scale, Distribution and Growth of Aviation: Implications for Climate Change' by Stefan Gössling & Andreas Humpe. *Global Environmental Change* Vol. 65, Nov 2020.

p. 33: Emissions from international flights do not count toward any one country's emissions, presenting few incentives to lower them. 'Climate Change and Flying: What Share of Global CO_2 Emissions Come from Aviation?' by Hannah Ritchie. *Our World in Data* 22 Oct 2020. Accessed 2 Jan 2022 https://ourworldindata.org/co2-emissions-from-aviation

p. 35: Natalie Diaz's poem 'The First Water Is the Body' can be found in *Postcolonial Love Poem*. London: Faber & Faber, 2020.

p. 38: George Monbiot argued that the only answer to aviation's catastrophic impact on global warming is to ground most aircraft in 'On the flight path to global meltdown'. *The Guardian* 21 Sep 2006. Accessed 2 Jan 2022 https://www.theguardian.com/environment/2006/sep/21/travelsenvironmentalimpact.ethicalliving

p. 39: Henry Shue's writing on emissions, privilege, and justice can be found in 'Subsistence Emissions and Luxury Emissions'. *Law & Policy* Vol. 15, No.1, Jan 1993.

THE GREAT MOOSE MIGRATION

p. 45: A video of highlights from SVT's *Den stora älgvandringen*, 2020, remains available to watch on svtplay.se in Jan 2022. The entire live stream from 2021 is available on the same website.

p. 45: The (now former) Swedish prime minister Stefan Löfven's statement about Swedish people acting responsibly was published on 6 Apr 2020 by the Prime Minister's Office. Accessed 2 Jan 2022 https://www.government.se/articles/2020/04/strategy-in-response-to-the-covid-19-pandemic/

p. 46: John Berger wrote about the dynamics of observing non-human animals in *Why Look at Animals*. New York: Penguin, 2009.

p. 52: Amber A'Lee Frost writes about the 'wild' in wild-life documentaries in 'The Viewing of Nature'. *Current Affairs* 7 Jun 2017. Accessed 2 Jan 2022 https://www.currentaffairs.org/2016/06/the-viewing-of-nature

p. 53: Jim Robbins' article on the relationship between ecocide and pandemics: 'The Ecology of Disease'. *The New York Times* 14 Jul 2012. Accessed 2 Jan 2022 https://www.nytimes.com/2012/07/15/sunday-review/the-ecology-of-disease.html

p. 56: The Swedish government allowed so-called 'leave traffic' of German soldiers on Swedish railways to occupied Norway in order to avoid war. Later, this would include war materiel. 'Sweden: Negotiated Neutrality' by Klas Åmark in *The Cambridge History of the Second World War* Vol. 2. Eds. Richard Bosworth & Joseph Maiolo. Cambridge: Cambridge University Press.

p. 56: About the Swedish National Institute for Race Biology at Uppsala University, in Swedish: *Allt som kan mätas är inte vetenskap: en populärhistorisk skrift on Rasbiologiska institutet* (All that can be measured is not science: a popular history of the Institute for Race Biology) by Lennart Lundmark. Stockholm: Forum för Levande Historia, 2007. Accessed 3 Jan 2022 https://www.levandehistoria.se/sites/default/files/material_file/skriftserie-4-allt-som-kan-matas-ar-inte-vetenskap.pdf

p.57: News article on moose and the threats of climate change published by Sweden's University of Agricultural Science: 'Klimatförändringar ett hot mot älgen'. 17 Apr 2022. Accessed 3 Jan 2022 https://www.slu.se/ew-nyheter/2020/4/klimatforandringar-ett-hot-mot-algen/

p. 57: The study confirming the spread of brain worm in Swedish moose: 'Epizootiology of *Elaphostrongylus alces* in Swedish Moose' by Margareta Stéen, Ing-Marie Olsson Ressner, Bodil Olsson & Erik Petersson, *Alces: A Journal Devoted to the Biology and Management of Moose* Vol. 52, 15 Aug 2016.

p. 57: Attenborough's comment about the series *Dynasties* and politics was made in an interview: 'David Attenborough: too much alarmism on environment a turn-off' by Jonathan Watts. *The Observer* 4 Nov 2018. Accessed 3 Jan 2022 https://www.theguardian.com/environment/2018/nov/04/attenborough-dynasties-ecological-campaign

A NATURALISATION

p. 71: 'Reinnervation Post-Heart Transplantation' by Avishay Grupper, Henry Gewirtz & Sudhir Kushwaha. *European Heart Journal* Vol. 39, No. 20, 21 May 2018.

p.71: 'Post-Transplant Nerve Regrowth Better with Young Hearts, Quick Surgery'. *ScienceDaily* 15 Jul 2002. Accessed 3 Jan 2022 https://www.sciencedaily.com/releases/2002/07/020715075935.htm

p. 72: Stefan Zweig wrote about travelling before WWI in *The World of Yesterday*, transl. from the German by Anthea Bell. London: Pushkin Press, 2014.

p. 76: The quoted poem by Pia Tafdrup is 'Pass, Passport, Passaporto, etc.' transl. from the Danish by David McDuff, published in *Salamander Sun & Other Poems*. Hexham, England: Bloodaxe Books, 2015.

p. 81: 'British citizenship of six million people could be jeopardised by Home Office plans' by Ben van der Merwe. *The New Statesman* 1 Dec 2021. Accessed 3 Jan 2022 https://www.newstatesman.com/politics/2021/12/exclusive-british-citizenship-of-six-million-people-could-be-jeopardised-by-home-office-plans

p. 81: Home Office news story: 'Government Cracking Down on Highly Disruptive Protest Tactics', 2 Dec 2021. Accessed 3 Jan 2022 https://www.gov.uk/government/news/government-cracking-down-on-highly-disruptive-protest-tactics

p. 81: The term Gypsy, Roma, and Traveller (GRT) communities is used officially in the UK by community-led organisations advocating for Gypsy, Roma and Traveller people. For community-led coverage of the impact of the proposed law, I recommend following Travellers' Times: http://www.travellerstimes.org.uk/

MIXED SIGNALS: FIVE MOMENTS OF UN-BELONGING

p. 92: 'The Pandemic Is a Portal' by Arundhati Roy in *Azadi: Freedom, Fascism, Fiction*. New York: Penguin, 2020.

OUT OF THE TUNNEL

p. 95: Adrienne Rich's words on the humanly possible, quoted in this essay, are found in 'Interview with Rachel Spence' as published in *Arts of the Possible: Essays and Conversations*. New York: Norton, 2002.

p. 97: *Corona, Climate, Chronic Emergency: War Communism in the Twenty-First Century* by Andreas Malm. London: Verso, 2020.

p. 98: 'In a Fight over a Colombian Coal Mine, Covid-19 Raises the Stakes' by Lise Josefsen Hermann'. *Undark* 22 Jul 2020. Accessed 3 Jan 2022 https://undark.org/2020/07/22/colombian-coal-mine-wayuu-covid-19/

p. 104: 'We Don't Have To Halt Climate Action To Fight Racism' by Mary Annaïse Heglar. *HuffPost* 12 Jun 2020. Accessed 3 Jan 2022 https://www.huffpost.com/entry/climate-crisis-racism-environmenal-justice_n_5ee072b9c5b6b9cbc7699c3d

p. 106: Writer Derek Thompson explores the idea of 'workism' in the article 'Workism Is Making Americans Miserable'. *The Atlantic* 24 Feb 2019. Accessed 3 Jan 2022 https://www.theatlantic.com/ideas/archive/2019/02/religion-workism-making-americans-miserable/583441/

p. 108: Ronald Melzack's 'neuromatrix' theory of pain is accessibly explained in Chapter 8, 'Chronic Pain', of *Pain: A Very Short Introduction* by Rob Boddice. Oxford: Oxford University Press, 2017.

p.109: The mentioned study is 'Cognitive Tunneling, Aircraft-Pilot Coupling Design Issues and Scenario Interpretation Under Stress in Recent Airline Accidents' by Meredith A. Bell, Eugenio L. Facci & Razia V. Nayeem. *International Symposium on Aviation Psychology*, 2005.

p. 111: Amin Maalouf's words on identity as a parchment can be found in *In the Name of Identity: Violence and the Need to Belong*, transl. by Barbara Bray. New York: Arcade Publishing, 2012.

p. 111: Audre Lorde's talk 'Learning from the 1960s' is included in *Your Silence Will Not Protect You*. London: Silver Press, 2017.

FREAK AGUACERO

Although no works are cited in this essay, one book above all was key to finding my way in to writing about Colombia for the first time in many years. *My Colombian War: A Journey Through the Country I Left Behind* by Silvana Paternostro (New York: Macmillan, 2008) doesn't deal with environmental degradation or climate collapse to any significant extent, but Paternostro's writing about belonging and privilege with regards to the Colombian armed conflict was helpful and I'm very grateful to it.

BIRTH STRIKE: A STORY IN ARGUMENTS

p. 133: The analogy of a crashing airplane was made by Veerabhadran Ramanathan to illustrate the findings of the report 'Well Below 2 °C: Mitigation Strategies for Avoiding Dangerous to Catastrophic Climate Changes', co-authored by Ramanathan & Yangyang Xu (published in the *Proceedings of the National Academy of Sciences* Vol. 114, No. 39, 6 Sep 2017). Ramanathan's statement can be found in the news article 'The Window Is Closing to Avoid Dangerous Global Warming' by Jean Chemnick, *Scientific American* 15 Sep 2017. Accessed 3 Jan 2022 https://www.scientificamerican.com/article/the-window-is-closing-to-avoid-dangerous-global-warming/

p. 140: The Oxfam report 'Confronting Carbon Inequality', from which these figures are taken, was published in Sep 2020. It can be downloaded from https://www.oxfam.org/en/research/confronting-carbon-inequality

p. 140: 'The outcry over ICE and hysterectomies, explained' by Nicole Narea, *Vox* 18 Sep 2020. Accessed 3 Jan 2022 https://www.vox.com/policy-and-politics/2020/9/15/21437805/whistleblower-hysterectomies-nurse-irwin-ice

p.141: 'Twice as many children waiting to be adopted as families to adopt'. *The Guardian* 14 Oct 2019. Accessed 3 Jan 2022 https://www.theguardian.com/society/2019/oct/14/twice-as-many-children-waiting-to-be-adopted-as-families-to-adopt

p. 142: *On Infertile Ground: Population Control and Women's Rights in the Era of Climate Change* by Jade S. Sasser (New York: New York University Press, 2018) was immensely helpful to Blythe and me when manoeuvring the development of Birth Strike and understanding the connotations for climate-justice work.

Jessica Gaitán Johannesson

ON WHETHER OR NOT TO THROW IN WHOSE TOWEL: A PERSONAL ENCYCLOPAEDIA OF HOPE

To learn about dystonia, the experiences of people living with the condition, and current research, please visit www.dystonia.org.uk

p. 160: I first came across the essay 'We Need Courage, Not Hope, to Face Climate Change' by Kate Marvel when it was quoted as part of a lecture. The essay was published on 1 Mar 2018 by the On Being Project. Accessed 3 Jan 2022 https://onbeing.org/blog/kate-marvel-we-need-courage-not-hope-to-face-climate-change/

p. 162: Rebecca Solnit's quoted lines are from the 2016 edition of *Hope in the Dark: Untold Histories, Wild Possibilities*. London: Canongate, 2016).

p. 162: Jan Zwicky's piece 'A Ship from Delos' can be found in the book *Learning to Die: Wisdom in the Age of Climate Crisis* by Robert Bringhurst & Jan Zwicky. Regina, SK: University of Regina Press, 2018.

p. 164: 'Uncivilisation', the Dark Mountain manifesto by Paul Kingsnorth and Dougald Hine remains accessible on the project's website.

p.164: The quoted statements about hope by Paul Kingsnorth can be found in an interview by Wen Stephenson, published in *Grist* 11 Apr 2012. Accessed 3 Jan 2022 https://grist.org/climate-energy/i-withdraw-a-talk-with-climate-defeatist-paul-kingsnorth/

p. 165: The documentary *De aarde draait door* (English title: *The Battle Against Climate Change by Paul Kingsnorth*) follows Kingsnorth's life in Ireland, and more recent thinking on climate and environmental collapse. It was broadcasted by VPRO in 2018 and is available on YouTube.

p. 167: *Active Hope: How to Face the Mess We're in Without Going Crazy* by Joanna Macy and Chris Johnstone. San Francisco, CA: New World Library, 2012.

p. 172: 'El derecho de soñar' by Eduardo Galeano. *El País* 26 Dec 1996. Accessed 3 Jan 2022 https://elpais.com/diario/1996/12/26/opinion/851554801_850215.html An English translation of the text, 'The Right to Dream' (quoted here), was published by *New Internationalist*. Accessed 3 Jan 2022 https://newint.org/blog/2015/04/13/galeano-right-to-dream